Landscapes
Ways of imagining the world

We work with leading authors to develop the
strongest educational materials in geographical sciences,
bringing cutting-edge thinking and best learning
practice to a global market.

Under a range of well-known imprints, including
Prentice Hall, we craft high quality print and
electronic publications which help readers to
understand and apply their content,
whether studying or at work.

To find out more about the complete range of our
publishing, please visit us on the World Wide Web at:
www.pearsoned.co.uk

Landscapes
Ways of imagining the world

Hilary Winchester, Lily Kong and Kevin Dunn

Harlow, England • London • New York • Boston • San Francisco • Toronto • Sydney • Singapore • Hong Kong
Tokyo • Seoul • Taipei • New Delhi • Cape Town • Madrid • Mexico City • Amsterdam • Munich • Paris • Milan

Pearson Education Limited
Edinburgh Gate
Harlow
Essex CM20 2JE
United Kingdom

and Associated Companies throughout the world

Visit us on the World Wide Web at:
www.pearsoned.co.uk

First published 2003

ISBN 978-0-582-28878-2

British Library Cataloguing-in-Publication Data
A catalogue record for this book is available from the British Library

10 9 8 7 6
11 10 09

Typeset in 11/12pt Adobe Garamond by 35
Printed and bound in Malaysia, VVP
The Publisher's policy is to use paper manufactured from sustainable forests.

Contents

List of figures

List of boxes

List of tables

Acknowledgements

We would like to thank numerous people who have assisted in the production of this book. A number of individuals provided research, editorial and secretarial assistance and we are particularly grateful to Betty Andrews, Lynette Boey, Justin de Rosa, Catherine Evans, Paul Foley, Kavita Gosavi, Bronwyn Hanna, Anton Kozlovic, Tan Guat Lee, Amy McDonald and Kamalini Ramdas for their contributions.

We would each like to thank our respective departments and institutions for their support. We also acknowledge the debt to our many students as much of this materials was initially drawn together as teaching materials for our courses in cultural geography. Lily Kong would like to acknowledge the financial support provided from NUS Academic Research Grant R–109–000–014–112. We have also valued the support of our families throughout this project.

We are grateful to the following for permission to reproduce copyright material:

Figure 1.1 from 'Scales, lines and minor geographies: whither King Island?' in *Australian Geographical Studies*, 37(3), 248–67, Blackwell Publishing, (Bradshaw, M. and Williams, S. 1999); Figure 2.1 from *A Geography of the Third World*, Routledge, Dickenson, J.P., Clarke, C.G., Gould, W.T.S., Prothero, R.M., Siddle, D.J., Smith C.T., Thomas-Hope, E.M. and Hodgkiss, A.G., 1983; Table 3.1 from *Getting the Picture: Essential Data on Australia Film Televison, Video and New Media*, Sydney, © Australian Film Commission, 1998; Figure 4.2 from 'Landscape structure of the Venetian myth' in *Journal of Historical Geography*, Vol. 8(2), Academic Press, Cosgrove, D., 1982; Figure 4.6 from Alan Moir, 1988; Figure 5.1 from *The social meanings of conflict in riots at the Australian Grand Prix motorcycle races*, Taylor & Francis Ltd, Cunneen C. and Lynch, R., 1988; Figure 5.2 from *Introducing Human Geography*, Pearson Education Australia, Waitt, G., McGuirk, P., Dunn, K.M., Hartig, K.V., and Burnley, I.H., 2000; Figure 6.4 from *Landscapes of Despair, from deinstitutionalization to home-lessness*, Oxford: Polity Press, Dear, M. and Wolch, J., 1987; Figure 7.1 from Bill Sheridan.

In some instances we have been unable to trace the owners of copyright material, and we would appreciate any information that would enable us to do so.

List of abbreviations

AFC	Australian Film Commission
BHP	Broken Hill Proprietary Company Limited
CCEA	The Central City East Association
CDC	Community Development Corporation
EPCOT	Experimental Prototype Community of Tomorrow
FACE	Families Against a Contaminated Environment
FIST	Friends Insist Stop Toxics
HDB	Housing and Development Board
	[This name is a guess based upon Page 113, Paragraph 1, Line 14]
HIFI	Hunter Industrial Facilities, Inc.
HIV	Human Immunodeficiency Virus
HIV-AIDS	Human Immunodeficiency Virus–Acquired Immune Deficiency Syndrome
LAPD	Los Angeles Police Department
LRC	Local Review Committee
LTTE	Liberation Tigers of Tamil Eelam (aka Tamil Tigers)
NIMBY	not in my backyard
NSS	The Nature Society of Singapore
NSW	New South Wales
PACE	People Against a Contaminated Environment
RCIADIC	Royal Commission into Aboriginal Deaths in Custody
STB	Singapore Tourism Board
TOADS	Temporary Obsolete Abandoned Derelict Sites
UNESCO	United Nations Educational, Scientific, and Cultural Organization
URA	Urban Redevelopment Authority
WTO	World Trade Organisation

Chapter 1

Cultures and landscapes

1.1 The desire for cultural landscapes

The discipline of geography has long maintained a focus on cultural land-scapes. Curiosity with landscapes and peoples continues to constitute an important impetus for geographical inquiry today. Populist and political sup-port for the discipline was founded on the 'discovery' of, and reporting back on, 'new' places and peoples. For example, governments in western countries have funded scientific expeditions to the so-called New World since the fifteenth century, and geographical societies have continued this tradition for the last 200 years. Geographical discovery and exploration facilitated and legitimated territorial claims and opened up new lands (Hartshorne, 1939:35–48; Powell, 1988), as an integral part of the colonial enterprise.

Geographers involved in all these expeditions produced texts about the cultures and landscapes they encountered and early cultural geographers were particularly interested in the diffusion of cultures. Geographical interest in cultural landscapes, of course, emerged even before the rise of western colonialist geography, in the form of explorers in the Middle Ages (see Box 1.1). Some of the texts were used to amuse and titillate an increasingly literate public with

---

Box 1.1   Capturing landscapes: the pre-European colonialists

It is generally assumed that the Middle Ages were largely typified by a lack of geography (Park, 1994:8). This conclusion ignores a rich vein of Islamic geography, particularly between the ninth and fourteenth centuries AD (Ahmad, 1947; Sharaf, 1963). The Islamic realm during this period con-tained great centres of learning, in which the Greco-Roman tradition was preserved, and from which large parts of Asia, Africa and Europe were mapped and described and new ideas in astrology, climatology and navigation devel-oped (Jones, 1993:13, 24, 26; Rosenthal, 1965:xvi; Sharaf, 1963:60). Islamic scholars developed Ptolemy's ideas on geography and astronomy (Rosenthal, 1965:214–23; Sharaf, 1963:83). They advanced ideas about erosion, landforms and the folding of mountains, many of which were counter to Christian creationism (Sharaf, 1963:84, 111). Symposia on meteorology, climatology

(continued)

1

(continued)

and other aspects of geography were held and the annual pilgrimage to Mecca 'itself fostered the exchange of geographical information' (Sharaf, 1963:70–1, 84). Al Maqdisi was a populariser of geography in the tenth century. He introduced the use of symbols and colours to cartography in an attempt to broaden academic interest in the discipline as well as to make the subject more accessible to the ordinary people (Sharaf, 1963:91–2).

Muslim explorers traversed much of south and central Asia, as well as the archipelagos and peninsulas of southeast Asia (Sharaf, 1963:76–8). Muslim traders were firmly established in China at least by the ninth century and Muslim armies had entered Sinkiang in 714 AD (El Erian, 1990:90). Islamic geographers ventured north through central Asia to the Volga region and Siberia, as well as through contemporary France, Germany and Denmark (Sharaf, 1963:74–5, 132). Muslim cartographers compiled ever greater maps and directories of the(ir) known world. The *Book of Routes and Kingdoms* (Ibn Khordathabak, 846) outlined all the principal routes of the Islamic world, including descriptions of passageways through central Asia and to the Chinese border and Silk Road. The islands of Sumatra and Borneo and other parts of what are now Malaysia and the Philippines were at least vaguely known to the Muslims during the tenth century AD (Ahmad, 1947:ii, 47; Sharaf, 1963:76–8, 141).

There is a colonialist arrogance and pervading sense of superiority in the Islamic geographers' depictions of African and European places on the edges of, or beyond, the Muslim realm. They described the economies and the 'backward cultures' they encountered. Ibn Jubeir, born in Valencia, made a scathing critique of the Bega tribes people of southeast Egypt. He was annoyed at what he saw as their cruelty and greed and that they went naked but for a loin cloth. Sharaf (1963:120–1) recorded that he considered them 'without mind, character or manners and, in his opinion, [they] should be cursed'. Like the Muslim explorers before him and the European colonials afterward the Tangier-born Ibn Battuta was by no means a cultural relativist. All the societies he encountered were judged by the standards he learned in the cultural centres of the Islamic realm. He described the town of Sayla in Somalia as the dirtiest and most boring on earth. He was particularly critical of societies in which women were unveiled and accorded too high a status (Sharaf, 1963:137, 143–4). Islam was practised in Iwalaten but he was irritated by the prejudice against white people that he encountered and he 'did not like their food' (Sharaf, 1963:144)! In the Maldive Islands he was appointed as a judge, but he soon made himself unpopular and had to leave. He had tried to compel the citizenry to follow a more 'civilised' way of life, insisting that they attend religious services at the mosque and compelling all women to wear clothes (Sharaf, 1963:140). Clearly, the great Islamic 'discoveries' and discoverers were as ethnocentric and as arrogant as the Christian explorers of European colonialism (although many centuries earlier). They also demonstrated the certainties and assuredness of modernist scholars.

tales of the exotic. More recent historic examples include Captain Richard Butler's search for the source of the Nile River in Africa and the Burke and Wills trek into Australia's interior. A contemporary manifestation would be the documentaries that result from expeditions funded by National Geographic, Discovery Channel and other media corporations. This book thus contributes to a long lineage of inquiry into cultural landscapes.

1.2 Defining 'culture' and 'landscape'

Two specific terms have a centrality in this book: 'culture' and 'landscape'. Most people would have an intuitive understanding of what is meant by these terms. Indeed, they are terms used in ordinary conversation by everyday people. But these seemingly straightforward terms have complex histories and multiple meanings. In Chapter 2, we will demonstrate how changing conceptions of both these terms have influenced the work of geographers. For now, we briefly review the definitions of 'culture' and then 'landscape'.

The term **culture** has come to have a range of meanings, many of which are inconsistent. In its classical use, culture was a synonym for civilisation. A cultured person was someone who had achieved a finery in their habits and tastes, such as appreciation of the fine arts of opera, theatre and literature. These pursuits were referred to as 'high culture'. In opposition to high culture was 'low culture' or the 'uncultured', including popular sports or mass-consumed television programmes. The behaviour associated with low culture was considered to be base or even savage. In this classical conception of culture, the common people, the working and middle classes, were deemed as culturally lesser than the elite. In this book, alongside our discussion of the extraordinary and the spectacular, we have an abiding interest in the ordinary cultures of everyday landscapes and we reject the portrayal of such culture as lesser or base. A second common definition is where culture is a synonym for ethnicity or the more problematic notion of race. Culture by this definition refers to distinct ethnic groups, categorised by language, religion, nation of birth, notions of race or indigeneity. This definition excludes gender, sexuality, class and other ways of life from analysis. Often, this definition of ethnicity was used to sort and categorise people. Indeed, race has been a particularly potent and problematic means for the categorisation of people. As we show in Chapter 2, the notion of racial groups was historically used to construct racial hierarchies. These hierarchies accumulated particular legitimacy – and therefore impact – from the supposed naturalness of those categories. According to this definition, culture was a sort of ethnic container into which people were born. Our view is that culture is much more dynamic and individualised than that definition allows and, further, that we are able to change our own culture and influence that of our children and peers.

In Chapter 2, we outline the debates surrounding this conception of culture within geography. Culture for us is best described as a 'way of life'. We imagine culture to be individually lived, dynamic and unique. At the same time, we recognise that culture is shared: it is a group phenomenon. Group affiliation and participation is one of the central means by which cultural groups are reproduced. Our central theoretical position, as articulated in the following chapter, is that culture is (re)produced – it is not 'natural'. Humankind are not born into static cultural groups that we cannot transcend. We hold culture to be socially constructed – a dynamic product of individuals and groups, both past and present (see also Chapter 2).

In this book, our focus is on **cultural landscapes**. Our interest is not so much with natural landscapes, if such landscapes even exist. Even those landscapes that are relatively unmodified (and therefore 'natural') are invested with cultural meaning through representations of them. But the landscapes that are modified and utilised by humans are those which most engage our geographical interest. In this book, we sometimes refer to the environment, sometimes singling out the built environment for attention. Environments are broader than specific landscapes and include associated biophysical and social contexts. Likewise, the term cultural landscape refers to more than just the surface of the earth. The cultural landscape includes buildings of various forms, such as houses, factories, monuments, barriers and so forth. At other times in this book we talk about places. Places are much more than cultural landscapes; they include the people and the relations between them. Places also comprise the images of those people and their landscapes; their senses of place and loyalties.

1.3 Moving through *Landscapes*

In reading this book, readers will be aided if they bear in mind that we have purposefully emphasised six themes. The degree of attention given to these themes varies within each chapter. We ask only that readers be cognisant of these themes as they traverse our examinations of landscapes.

1.3.1 Multiple cultures and landscapes

In *Landscapes* we explore a vast array of ways of life. **Ethnicity** is a powerful influence on groups and individuals and we examine the impact on the landscape of ethnic and relatedly, religious groups in a variety of local and national contexts, ranging from groups who are culturally dominant to others which work doggedly to assert their presence. Identity and group affiliation occurs through the landscape and this dynamic construction of culture is often perceptible. Ways of life are profoundly affected by gender relations and gendered constructions, and hence we look at **masculinities and femininities**, including the landscape impacts of dominant heterosexuality, notably as it affects and is resisted by gays and lesbians. *Landscapes* is also profoundly imbued with notions of **class**, including the distinct and related cultural landscapes of the elite and the marginalised poor. Our discussions of culture are often about hierarchies of power, in which some groups have become privileged as compared to others. The subjective and fluid nature of cultures does not discount the rise of new dominant and oppressive cultural forms and their spatial reflections. Such hierarchies span the axes of 'race', gender, sexuality, religion and class. Many texts have discussed the issues of racism, sexism, homophobia and elitism that underpin these hierarchies. In *Landscapes* we show how such politics are embodied, reflected and reinforced by the cultural landscape.

1.3.2 Landscapes and power

Explorers of the Middle Ages (see Box 1.1) provided the detailed descriptions of the places they 'discovered' (Sharaf, 1963:75–6). These records often involved criticism, of cultures in India, China and elsewhere (Ahmad, 1947:133–53; Sharaf, 1963:77, 111, 139). These peoples and their cultural landscapes were judged as either civilised or heathen, from the ethnocentric standpoint of the colonial geographer (see Box 1.1).

Of course, these 'new' peoples occupied land with which they were entirely familiar. The land had been discovered long ago by their ancestors. Indigenous Australians had encountered the Australian landmass at least 60,000 years before Captain Cook of the Royal British Navy and the earlier Dutch explorers (Flood, 1995). But to the colonial mind, these places came into existence when they became known within the 'civilised thinking' of Islamic or western society. The colonialism underlying geographical interest in cultural landscapes brings to the fore the issue of power. Power, control and resistance, are core foci of our discussions about landscapes. Cultural landscapes are part of a process in which hierarchies are reproduced and challenged.

1.3.3 Landscapes and texts

We are aware of many more places in the world, many more landscapes, than we have actually visited (Zonn, 1984:35). This awareness comes from texts, which take many forms. For example, feature films, television programmes and newspapers are fundamentally involved in mediating information about cultural landscapes (Burgess, 1985; Dunn and Winchester, 1999). Our individual stocks of such information can be as much informed by representations of cultural landscapes as they are by direct experience. Knowledge about cultural landscapes is produced through direct experience (tourism, trekking, fieldwork, passing through) as well as consumed from media (documentaries, travel brochures, lifestyle television) (see Zonn, 1984:36). The media and other mass cultural products are central to the generation of group identities and can be forceful tools in oppressions such as cultural imperialism (Said, 1978). Geographers have examined who is represented within media about places and who is excluded, how issues and interests are portrayed and the roles of media in political matters, such as contested notions of national identity (McManus and Pritchard, 2000:384). The gathering of geographical knowledge, and its conversion into texts, is central to the transmission of information about and (re)construction of cultural landscapes. For this reason, our discussions necessitate an inclusion of the texts on, or representations of, landscapes.

Representations of landscapes that have been produced by geographers are also texts. The records made by colonial geographers, and their governmental and populist consumption, demonstrate the importance of representations of landscape. The resultant texts included maps as well as 'scientific illustrations' of people and landscapes. These texts retain an official legitimacy and an enhanced authority, particularly as compared to most media or fictional representations. In the past, geographers have asserted the scientific or objective status of the representations they generated. The most powerful and apparently objective of these assertions is in the form of maps (Box 1.2). While overt claims to objectivity by geographers are less common in the contemporary era, it is still widely presumed that geographical knowledges are representative

---

Box 1.2   Maps as political texts

Maps may be seen as part of the colonial project. A recent study of the little known King Island in the Bass Strait between Tasmania and Victoria, Australia, emphasised that its mapping by British and French explorers in the early nineteenth century was the starting point of its colonial record as 'cartographic moments of European imperial territorialisation' (Bradshaw and Williams, 1999). The official cartography of the island normally relocated King Island closer to Tasmania and erased its relationship to Victoria (Figure 1.1). Bradshaw and Williams (1999:256–7) argued cogently that:

> This major cartography promotes misconceptions about the island's proximity to Tasmania and gives no indication of its possible relations with Victoria. As geographers, map-makers, writers of this form of discursive diagram, we like to think that we are alert to the sleights of hand that give maps their power. . . . Although technically aware of the cartographic treatment of King Island, we were largely unaware of the effects that even this obvious a dislocation had on our (mis)understanding of actual geographic, socio-economic and cultural relations on the island.

The capacity for distortion is clearly demonstrated in this example. Such distortion can occur with deliberate intent, as with the maps produced by marketeers that show the location they are selling to be at the centre of the known universe. Overt manipulation has resulted in the discrediting of some maps as official documents, particularly those maps produced by Nazi geopoliticians in the 1930s. World regional maps also served geopolitical aims of sorting imperial and colonised peoples. As Agnew and Corbridge (1995:58–9) pointed out, 'such mappings did express in extreme form the common assumption that the world was constituted of racial groupings that could be neatly divided into two "types" of people'. Maps have reinforced the idea of a hierarchy of racial groups. In other cases, different cartographic representations may give rise to conflict and in exceptional cases they have been used to counteract western-centrism (see Peters Projection, in Chapter 2).

(continued)

(continued)

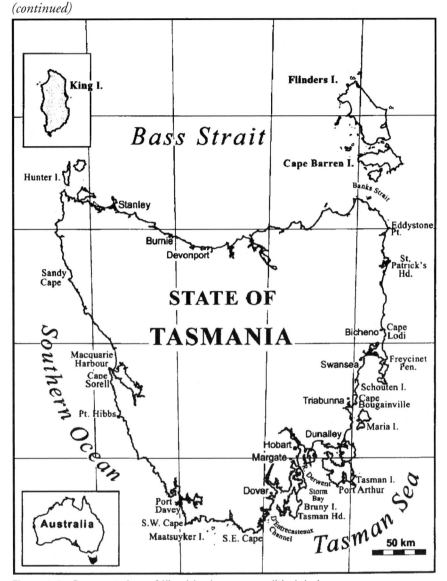

Figure 1.1 Representations of King Island, maps as political devices.
Source: Bradshaw and Williams, 1999:256 (Figure 3).

encapsulations of actual landscapes. However, such knowledge is still an attempt to re-present a landscape; it is a social construction and cannot transcend its representational status (Walton, 1995). In the next chapter, many of the different approaches to constructing knowledge about cultural landscapes are reviewed.

1.3.4 Landscapes and ordinary life

Too often, geographers have overwhelmingly focused their attention on cultures and landscapes thought to be extraordinary. Like the early anthropologists, geographers were most interested in cultures they considered to be different, such as ethnic minorities within cities or indigenous groups in other countries. This was partly a product of the western-centred nature of geography and the dominant thinking that saw non-western societies as objects of curiosity or problems. This focus on Other cultures, rather than the Self, is still strong within geography (Kong, 2001) and anthropology. In Chapter 3, we outline how geographers have reconsidered this emphasis on the exotic and extraordinary. In *Landscapes* we endeavour to examine the ordinary, although that is not always possible – neither is it necessarily desirable to do this exclusively. Nonetheless, we look to popular culture (for example, food, music and feature films) and we critically read day-to-day landscapes such as foreshore redevelopment areas and various aspects of domestic and suburban life such as street signs, fast food restaurants and housing. We even think about bodies and how they negotiate and perform in landscapes and of how they may even be landscapes in themselves. Landscapes that you might not consider to be ordinary are also covered. Ethnic precincts, so-called Chinatowns and ghettoes, world's fairs and so forth are important landscapes and they receive attention in *Landscapes*, as do places of worship and landscapes of protracted inter-communal conflict.

1.3.5 Landscapes of the Pacific Rim

In the writing of *Landscapes*, we bring a slightly different view from that of most texts written in the area of cultural geography. Even while we refer to the works of cultural geographers from Europe and North America, we have attempted where possible to relate those analyses to the circumstances of the Pacific Rim and, especially, Australia and Asia. For example, we look at ethnic concentration and cultural landscapes in such dramatically different contexts as Sydney and Singapore, Sri Lanka and Indonesia. The varying political and cultural contexts of each city are interesting in themselves, but each case study helps us better understand the way that cultural landscapes are constructed in unique and particular ways, tempered by local cultural contexts, power relations and the existing built environment. In this, we are little different from geographers across the world. Nonetheless, we have endeavoured, wherever possible, to use examples from our own home countries and near neighbours. The analysis of Asia–Pacific situations does generate different findings and fresh insights into the links between culture and landscape. It is our earnest hope that *Landscapes* provides such insights.

1.3.6 Landscapes and scale

We have already introduced our argument that the ordinary has for too long been a neglected focus of analysis in geography. However, geographers have always retained a commendable interest in the local. Indeed, this focus on the local has at times drawn cultural geographers to the study of everyday landscapes. In *Landscapes*, we continue a traditional emphasis on the local, a rich site for analyses. At the local level, the opinions expressed are often unguarded, the fears and biases are palpable, the case at hand ephemeral and the material impacts immanent. The local is also an instructive site at which to examine issues of nationalism and globalisation. Our analyses range from the way landscapes figure in very localised politics, to national debates and to international issues of cultural globalisation. In *Landscapes*, we move between scales, from the international to the national to the local to the body and illustrate the intersections between them. But in all of this we retain an emphasis on actual landscapes and representations of them. In this sense, our discussions of nationalism, globalisation and regionalism are all grounded locally.

In the next chapter, we outline the changing ways in which geographers have analysed cultural landscapes. There is a rich history of landscape analysis. Formal analyses began with emphasis on how physical landscapes influenced (or even determined) cultures. Later, the emphasis was on how cultures were imprinted onto landscapes. In contemporary times, there is a focus on the textual representations of landscapes and a fundamental interest in how power relations are embodied within landscapes. In Chapter 3, we demonstrate how culture and cultural relations are reflected in ordinary landscapes of popular culture and we use as some examples everyday cultures of cuisine, music and film. In Chapters 4 and 5, we categorise the forms of politics that operate within and through landscape. These range from fundamentally oppressive regimes such as apartheid and compulsory segregation, to the separation of gendered spheres of activity to the issue of enforced heterosexuality (Chapter 4). We outline how landscapes are involved in the reproduction of power relations between cultural groups. In *Landscapes*, we also discuss the resistance and opposition to oppression that occur through cultural landscapes (Chapter 5). In Chapter 6, we look at marginal and liminal landscapes. Such landscapes facilitate change and transition, but can also consolidate exclusion. Finally, we journey to landscapes that are closest in. Bodies must negotiate landscape and, depending on the body, access and mobility are dramatically uneven. We entertain the idea that the body itself is a landscape, open to inscription and a means for expression. The body may be the largest scale of geographical analysis.

In *Landscapes*, the diverse and often powerful roles of cultural landscapes are demonstrated. Our hope is that those who journey through *Landscapes* will become more perceptive and empowered readers of the everyday but also extraordinary cultural landscapes they move through.

Bon voyage.

Chapter 2

Changing geographical approaches to cultural landscapes

The formal geographical study of cultural landscapes has changed quite dramatically over the last century. This change has been linked to a changing conception of culture and of its relation to landscape. At the start of the twentieth century, many geographers argued that landscapes, as part of the biophysical environment generally, determined the nature of the cultures which existed within them. This brand of geographical understanding has been referred to as environmental determinism, which later developed into environmental possibilism. From the 1920s, a competing approach considered landscapes to be simple repositories onto which cultures were deposited, in this way forming the cultural landscape. This second view of cultural landscapes has been associated with the Berkeley School of cultural geography. The Berkeley approach held cultures to be distinct and static entities and these were impressed on the physical landscape. In the 1990s, students of cultural landscapes have come to see culture as a process. To 'new cultural geographers', the cultural landscape is both an outcome and a medium of culture. Culture exists with*in*, as opposed to *on*, the landscape.

By way of example, these three general phases are perceptible in the development of geographical studies of religion over the century. Kong (1990:358–9) argued there has been a thesis–antithesis–synthesis series of shifts in the geographical study of religion:

> In the primary state of development [of the geography of religion], the focus was on a one-sided presentation of religion as determined by its environment; environmental explanations were sought, appropriately or otherwise, to aid the understanding of the origin of religions and religious practices. In the second stage of antithesis, the geography of religion moved to a one-sided study of the moulding influence of religion on its environment, to the point of shaping the settlement and landscape. . . . the field has clearly entered a third stage of development – synthesis. . . . a synthetic approach that focuses on the reciprocal network of relations between religion and environment.

In the following three sections, works within cultural geography and cognate disciplines are used to demonstrate the abovementioned shifts in the study of cultural landscapes. In sequence, these are, first, environmental determinism, then through the Berkeley School of cultural determinism, finally to the new cultural geographies.

2.1 Environmental determinism

In the earlier part of the twentieth century, geographers explored how different environments generated differing cultures. In these works, geographers
would demonstrate how the environment (including landscapes) was an independent variable or agent that determined culture. As the environment varied,
so would cultures. Perhaps the most well-known treatise on environmental
determinism was Ellen Semple's *Influences of Geographic Environment*. Semple
(1911:1) made the argument plain: 'Man [sic] is a product of the earth's
surface.' Similarly, one of the founders of Australian geography, Griffith Taylor
(1940:424–48), asserted that geography was able to demonstrate how humans
were conditioned by their environment. Indeed, Griffith Taylor carried environmental determinism into the 1950s in Australia, long after the Berkeley
School approach had diffused throughout North American geography.

Climate was shown by geographers to have influenced the myths and rituals
of religions. Ellen Semple (1911:41) observed that in Eskimo culture, Hell
was a place of extreme cold, whereas both Muslim and Christian sacred texts
depict Hell as comprising fire and boiling water and Heaven as a garden. In
the Qur'an, Heaven is described as a 'Garden underneath which rivers flow'
(see V:85; XIII:35; XLVII:15). The climate of Hell for Christians is mostly an
inferno, although the maps of Hell made by Dante, an Italian poet and commentator of the fourteenth century, also included sites of extreme cold (see
Astarotte, 2001:23–4). These associations between the afterlife and climate
clearly reflect the various environmental conditions that pertained in the cultural hearths of Christianity and Islam.

In many belief systems there has been a veneration of nature, or aspects of
the physical landscape, such as elevated places, caves, forests and trees and
heavenly bodies (Jordan, 1973:150–2). Blagg (1986:15) informed us that in
Roman Britain 'various natural features of the landscape had divine associations, either as the homes of gods or as gods in themselves'. These included
various river gods and the woodland god Silvanus. Hsu (1969) detected a
direct relation between natural calamities and religious belief in China, noting
the link between locust plagues and the worship of locust-gods. Furthermore,
there are many animistic religions that believe inanimate objects possess souls
or certain gods and should therefore be revered (Park, 1994:37). According to
many Australian indigenous faiths, various species of fauna and inanimate
objects are present-day incarnations of spirits. Some prominent landforms,
such as outcrops and mountains, are believed to be the contemporary manifestations of significant creation beings, often held responsible for developing
the surrounding landscape (Young, 1993). For example, Jabreen, or Burleigh
Mountain, in the Gold Coast of Australia, is the home of Jellurgal, the Dream-
ing God (Waitt *et al.*, 2000:167). Similarly, the contested redevelopment site
of the old Swan Brewery in Perth was an Aboriginal sacred site where Waugal,
a great serpent-like creation being, had burrowed into the earth to rest (Jones,
1997:134–7).

2.1.1 Environmental hierarchies

Operating from a European viewpoint, the geographer Jordan (1973:150) commented that the 'tie between religion and the physical surroundings is generally stronger among primitive peoples'. This reflected a colonial arrogance. Park (1994:27–9) referred to how this 'presuppositional hierarchy' had been utilised by geographers of religion. According to this structure, Christianity reigned supreme at the top of the civilisation ranking, animism and other 'tribal religions' were positioned at the lower end and religions like Judaism and Islam were accorded middling positions. Through texts and scientific papers, western geography helped reconstruct this religious hierarchy. Geographers also reproduced and developed racial and environmental hierarchies. Griffith Taylor, for example, produced racial categories of people and made links between environment and variations in skull morphology, stature and nose shape. Taylor (1949:51–67) drew links between degrees of skin colour and average temperatures, mapped 'zones of pigmentation' and examined the geography of skull morphology. His observations of body shape preferences reveal his ethnocentrism, as well as other hierarchies:

> For instance, slight figures are more popular among English folk, while plump figures are appreciated among the Southern Europeans, and very fat women are attractive to most Negroes. Blonds and Brunettes vary in favour (Taylor, 1949:57).

Taylor's argument was that some cultural traits, in this instance body shape preferences, were culturally contingent. In this regard, he was in accord with the contemporary geographic analyses of corporeality (see Chapter 7). However, Taylor's examples reveal a series of underlying hierarchies. There is a gender hierarchy in which women are always the sexual objects and men are implicitly the active sexual subjects. There is, of course, a 'racial' and ethnic hierarchy and this is paralleled by a body shape hierarchy in which slim is an Anglo-Celtic preference and 'very fat' is a negro preference. It was through the reproduction of such hierarchies that geographers reflected and reproduced colonialism.

In the racial hierarchies produced in the late 1800s and early 1900s, indigenous peoples were seen as (too) close to nature and therefore less cultured or civilised. People not so closely tied to their environment, such as Europeans who had 'mastered' their environment, were seen as superior. This hierarchy was used as a justification for the dispossession, extinction or exploitation of indigenous people by the supposedly superior 'races' (see Box 2.1). This reminds us again of the political implications and power of geographical explanation.

Geographers like Semple (1911) argued that it was environment which gave rise to 'races' of human beings. These categories were related to what have been called 'natural regions'. Natural regions were areas of coherent biophysical conditions and these regions were thought to give rise to specific ways of life (Herbertson, 1905:301). Clear examples of this thinking are seen in the 'environmental challenge-response' theories which were popular in the

---

Box 2.1   Racial hierarchy and the colonial expedition

Muslim Afghan cameleers were involved, with little recognition, in some of the foundational and even iconic aspects of Australia's European development. Many of the most well-known expeditions into the Australian inland included British explorers but also Afghan cameleers and indigenous guides. These included the Burke and Wills expedition, as well as infrastructure projects such as the construction of the overland telegraph.

One successful expedition in 1874 charted a route from South Australia to Western Australia. The expedition party was rewarded by the South Australian Government with a grant of £1000. The distribution of these funds demonstrates both the class and racial hierarchies operating in Australia at that time.

The expedition leader Colonel Egerton Warburton retained £500 and £200 went to each of the second-in-command and to Warburton's son. The white cook received £40. The two Afghan cameleers received £25 each and the Aboriginal guide got the remaining £10 (Rajkowski, 1987:27).

---

first few decades of the twentieth century and which have found favour again more recently. The thesis held that peoples who resided in more temperate and cold climates were compelled to be more innovative in satisfying their needs for food, warm clothing and accommodation, whereas people in tropical climates were sapped of the stimulatory need for creativity (Smith, 1913:7). Human development – advances in the technologies of production, clothing and shelter – was equated with civilisation. Equatorial climates were not therefore thought to be conducive to the development of civilised peoples or cultures. Of course, this thesis fails to provide a satisfactory answer to some critical questions. How did Hong Kong or Singapore come to be economic powerhouses? Why did not the colder regions of North America and most of Asia and Europe become early sites of technological advance (Diamond, 1997:22)? And why were tropical parts of South America the sites of some of the earliest forms of writing?

Nonetheless, it is clear that there have been differential rates of human development and there are compelling arguments that environment has had an influence on this variation. Jared Diamond (1997), a bio-geographer, has argued recently that differential development over the last 13,000 years has been linked quite overtly to environmental circumstances. Interestingly, his thesis is in part a critique of the notion that 'racial' differences explain differential development and the uptakes of new technologies (writing, political organisation, weaponry, etc.). Most white people still assume that the non-industrialised nature of indigenous Australian culture was associated with a primitiveness inherent to those people. And yet, all sound research and genetic evidence refutes the notion that there are definitive 'races' and that there are any links between genetic make-up and intelligence (see Human Genome Diversity Committee, 1993; UNESCO, 1983, etc.). Diamond argued that:

Box 2.2   Mega-fauna and the politics of indigenous impact in Australia

Jared Diamond argued that the spatially varied presence of animals available for domestication can 'go a long way toward explaining why only a few areas became independent centers of food production, and why it arose earlier in some of those areas than in others' (Diamond, 1997:29). Food surpluses from domestication and agriculture freed up labour for the development of technologies, industrialisation and so forth. It is thought that one of the reasons indigenous Australians did not develop such technologies was the lack of fauna suited to domestication.

The large mammals of Australia appear to have disappeared soon after humans arrived. It is thought by some that these fauna were naïve of the threat humans posed. They were quickly wiped out as humans spread through the continent in what has been termed an 'overkill' (Flannery, 1994). This overkill robbed Australia of those animals that could have become domesticated livestock and thus fundamentally influenced the way of life in indigenous Australia (Diamond, 1997:42–4). Critics of the overkill thesis have argued that it was more likely that climate changes were responsible for the cataclysmic demise of the Australian mega-fauna.

The thesis of overkill is a double-edged sword for contemporary indigenous politics, conflicting with the image of indigenous people as being 'at peace with the land', as being 'of the land' and spiritually connected to it (Hamilton, 1990:22). This image has been one of the few positive portrayals of indigenous Australians. They have suffered a demonstrably poor depiction of themselves and their spaces in the media and other cultural products (Meadows and Oldham, 1991; Royal Commission into Aboriginal Deaths in Custody, 1991:184–7). However, a considerable environmental impact, such as overkill, strengthens the sense of indigenous possession of and influence on the Australian landmass prior to European invasion. So does the impact of fire-stick farming regimes on flora and land management (Jones, 1969). Such impacts have been important in the context of contemporary land title debates, in which some Europeans have argued that the Australian landmass was *terra nullius* (effectively unoccupied and unmanaged) prior to European presence. Significant environmental impacts confound the thesis of *terrra nullius*.

> History followed different courses for different peoples because of differences among peoples' environments, not because of biological differences among peoples themselves (Diamond, 1997:25).

Diamond (1997:29–30) pointed to the important biophysical influences on the ability to cultivate agricultural products and the varied presence of large fauna that could be domesticated (see Box 2.2). These environmental conditions ultimately determined the ability of peoples to develop: to produce food surpluses and to free up labour for administration and invention. To talk of 'development' in this way is itself problematic, as it assumes that indigenous ways of life are less developed than those of people in the industrialised west and east (Rostow, 1960). It is more appropriate, and less ethnocentric, to talk of different ways of life, rather than differential rates of development. Clearly however, environment can have a fundamental impact on culture.

The maps and textual analyses of geographers have generally reinforced the ethnocentric division of the world into poor-uncivilised and rich-advanced realms. The standard Mercator Projection is thought to give prominence to developed western countries by exaggerating the size of the landmasses in the upper northern part of the globe. The more recently devised Peters Projection provides a less Eurocentric depiction in which the more densely settled parts of the planet are better represented (Figure 2.1).

2.1.2   The role of agency: environmental possibilism

The earlier enthusiasm by geographers to demonstrate how environments determined culture was largely abandoned by the latter half of the twentieth century. For example, Gay (1971:18) stated that '[f]ew geographers are now concerned with examining the effect of the environment on religion'. Nonetheless, there is little doubt that the environment has been, and remains, a significant effect on culture. Taylor (1940:409–13) demonstrated that population expansion in Australia would be severely limited by the very low annual rainfalls received throughout much of the country. Taylor's views were seen as unpatriotic as they contradicted national policy for rapid population expansion and he was hounded out of the country as a result (Powell, 1988; 1993). Geographical knowledges are powerful texts and can be threats to other powerful interests.

Diamond (1997:26) asserted that 'geography obviously has some effect on history; the open question concerns how much effect'. While there is now general acceptance that environment clearly *influences* culture, geographers are unwilling to restrict their analyses to a research framework focused on the ways in which environment *determines* culture. Environmental possibilism is an approach that recognises a range of possible cultural directions facilitated by the environment and that individuals retain a fair degree of agency in determining that direction. Finally, while culture is influenced by the physical environment, it is also clear that cultures are continually (re)produced by the social environment. As we show later, the specific dynamics of place feed into culture in a multitude of ways.

2.2 The Berkeley School of cultural geography

To the environmental determinists, culture could be said to be drawn up through the soles of your boots. But to Berkeley School geographers, notably Carl Sauer, culture was left as an imprint on the landscape. According to the former perspective, ways of life derive from the environment that surrounds people. But to the latter, cultures are largely deposited on the landscape.

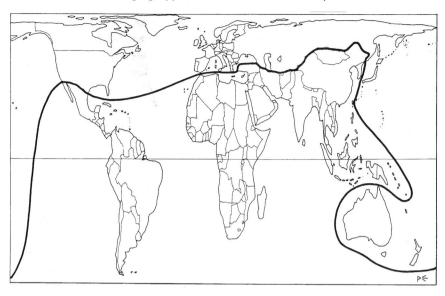

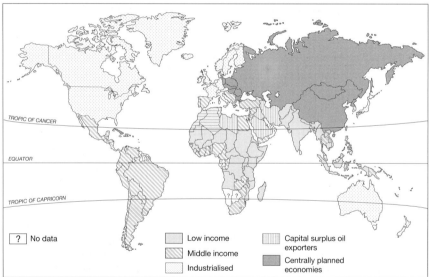

Figure 2.1 Contested depictions of the planet: rich and poor worlds. The Peters Projection (above) versus the Mercator Projection (below)

Source: J.P. Dickenson *et al.*, *A Geography of the Third World*.

The Berkeley School of cultural geography was led by one of geography's most influential figures, Carl Sauer (1889–1975). Sauer was Head of Geography at Berkeley from the 1920s to the 1950s (Jackson, 1989:10–13). His influence spread throughout North American geography and beyond (Price and Lewis, 1993:5–6). The Berkeley approach diffused from the University of California at Berkeley, USA, to influence geographers in other countries. In the late 1980s, this Berkeley tradition was given the dubious title of 'old cultural geography' (see Cosgrove and Jackson, 1987:95). This title implicitly recognised the prominent position that the Berkeley School had achieved within geography, but it also had the effect of rendering invisible the contributions of earlier geographies of the cultural landscape, such as the work of the environmental determinists.

A strong theme in the work of Sauer and his contemporaries was that a cultural landscape gave expression to the ways of life in a place. His most famous comment along this line was that:

> Culture is the agent, the natural area is the medium, the cultural landscape is the result (Sauer, 1925:46).

This statement reflects what many have thought to be a purposeful counter to environmental determinism. In Sauer's thinking, it was culture that had agency rather than the environment (see Kong, 1997:183). Sauer's work aimed to map the distribution and dispersion of cultures across space. He was interested in how culture spread throughout regions, how some cultural traits displaced others and the relationship between the natural landscape and culture.

'Old cultural geographers' tended to focus on the ways in which culture altered the physical landscape. For example, geographers of religion mainly restricted their analyses to the spatial results of belief systems and they shied away from any examination of religion itself (Deffontaines, 1953:37; Fickeler, 1962:95, 117; Gay, 1971:18; Isaac, 1959–60:18; Sopher, 1967:1, 24). These geographies went so far as to identify the expected landscape impacts of cultural groups. For example, Rapoport (1984) was able to spell out the likely 'urban orders' that would result from particular cultural groups. Mexican, Roman, traditional Chinese and Muslim cities could be expected to have specific locations, relations of landscape, spacing, toponyms, orientations, colours, smells and activities (Rapoport, 1984:54). For example:

> In the Moslem city, whether in North Africa, the Middle East or North India, one also finds the bazaar as a linear element (as opposed to the node markets – market squares – of Europe). Behind and between these linear bazaars, one finds the residential communities described above. These named quarters, and very small tightly knit subareas of clans, large families, occupational or common origin groups, may only be accessible through culs-de-sac off the bazaar (Rapoport, 1984:58–9).

The generalisations within Rapoport's description of Muslim cities is a marker of Berkeley-style confidence that cultures were distinct, static, and therefore predictable.

Much of Berkeley-style cultural geography focused on vernacular landscapes. American cultural geography in the Berkeley tradition produced exhaustive surveys and atlases of the forms and materials of houses, barns and fences (see Fellman *et al.*, 1995:213–29). European geography from the 1930s also followed a similar, yet slightly disparate, direction. This specific deviation owed much to the traditions of regional geography, which is mainly characterised by the belief that areas share particular and distinct cultural traits (see Box 2.3).

Box 2.3   The *Landschaft* School: expanding the definition of landscape

The Landscape School of geography developed from diverse origins predominantly from European traditions. The interwar period of geographical thought was heavily influenced by the German concept of *Landschaft* (generally referring to a bounded area) and the Dutch idea of *landschap* (used primarily in a visual and artistic sense). It also derives from the very strong regional geography of the French school of almost a century ago. The key exponent of this style, Paul Vidal de la Blache, wrote his acclaimed work *France de l'Est* in 1917. This work presented a finely grained chronological development of rural landscapes and societies over two millennia. This view of landscapes was that they were essentially organic, growing and developing in symbiotic relationship with their societies. This organic view of landscapes is rather different in emphasis from that of Sauer and the Berkeley School where the 'natural' landscape was seen as a blank sheet or palimpsest to be overprinted with the impact of human activity.

Landscape, even in the interwar period, was a problematic term. Hartshorne (1939:149–74) devoted a whole chapter to its definition in his work *The Nature of Geography*. He concluded that a landscape was not just a region or a bounded area, although the terms had often been used interchangeably. He felt that the term landscape was most appropriately used to encompass the visible and material objects on the earth's surface, while scorning any view of the landscape as symbolic or, as he termed it, 'the concept of "landscape" as reservations' (Hartshorne, 1939:168).

Hartshorne therefore, as an American, supported the Berkeley view of landscape as a cultural landscape, whereas most European landscape geography of the period would probably now be defined as regional geography. This European view of landscape as region, whereby the study of landscape and society provided geography with its own distinctive subject matter and method, was seen as a defining strength of geography. Dickinson (1939:8) considered that the identification of 'a definite body of material for investigation (represented) the most significant and widespread trend of post-war geography'.

The complex history and usage of the term means that, inevitably, landscape now has multiple meanings and interpretations as a region or place, as a collection of artefacts and as a representation either in another medium (on canvas, in film) or as a representation of culture through symbolic means.

Critics of 'old cultural geography' have argued that the Berkeley School was severely limited. The main criticisms focus on its superorganic conception of culture (Duncan, 1980; Jackson, 1989:16–9), its atheoretical and artefactual approach (Duncan, 1990:3, 11–3; Jackson, 1989:11, 19) and its rustic or exotic empirical focus (Cosgrove, 1989:570).

2.2.1 Critique 1: the superorganic conception of culture

In moving beyond environmental determinism, and by foregrounding the role of culture, it may be said that Sauer and his contemporaries replaced environmental determinism with a cultural determinism. Not only did culture determine landscape, but culture also determined the individual. This determinism has been referred to as a superorganic conception of culture. The superorganic conception of culture within the Berkeley School has been traced to the influence of anthropologists such as Kroeber, Boas and Lowie (Duncan, 1980; Jackson, 1989:16–19).

Put simply, a superorganic approach adopts the view that culture is an entity at a higher level than the individual, that it is governed by a logic of its own and that it actively enables and constrains human behaviour (Duncan and Ley, 1993:11; Jackson, 1989:18). The work of Sauer (1925) and Zelinsky (1973:40–1) are said to have ascribed a superorganic status to culture (Jackson, 1989:17–20):

> Culture can be regarded as the structured, traditional set of patterns for behaviour, a code or template for ideas and acts . . . it appears to be a superorganic entity living and changing according to a still obscure set of internal laws. Although individual minds are needed to sustain it, by some remarkable process culture also lives on its own, quite apart from the single person (Zelinsky, 1973:70–1).

In similar vein, Rapoport (1984:51) explained that culture was 'about a group of people who have a set of values and beliefs which embody ideals and are transmitted to members of the group through enculturation'. From this perspective, people are born into a culture or they quickly adopt a culture from their family and peers. The superorganic conception also sets culture apart from the environment – there is no room for discussing environmental influence. Culture as a totality was imprinted on people and through them the landscape. Sauer and his contemporaries categorised cultures and then mapped them to explain their spatial distribution and impacts. And while these geographers had a great interest in the diffusion of a culture, often mapping the migrations of cultural groups, they would have had much less of a focus on the manner in which cultures changed through the migration process. In a superorganic conception of culture, the idea of 'individual culture' was precluded, ignoring the internal heterogeneity within a cultural group. It precluded analysis of contested notions of culture and generally denied the possibility of cultural transformation.

However, long before the rise of the 'new cultural geography' in the 1990s, individual researchers were moving beyond the confines of superorganicism. As we show later, cultural geographers were identifying the multiplicity of meanings in the cultural landscape, the socially constructed nature of culture and the contested nature of landscape interpretation.

2.2.2 Critique 2: artefactual, object fetishism

The charge of artefactuality is one of the key critiques of 'old cultural geography'. The alleged limitation to surveys or catalogues of landscape artefacts has been described as 'object fetishism' (Duncan, 1990:11; Goss, 1988:393; Jackson, 1989:19). The old approaches were considered to have an empiricist fixation with the physical aspects of culture. Duncan and Ley (1993:11) observed that much of this work 'continued on in the tradition of material cultural studies, observing, describing, classifying and mapping such cultural artefacts as houses, barns or fences in order to identify cultural regions or paths of cultural transmission'. These included analyses that calculated the number of log houses built to a certain design in a region or the mapping of the sites of churches throughout a city.

As an example, the geographies of religion were often summaries of impacts that religion had on the landscape. Sopher's (1967:24–34) seminal work on the geographical impacts of religion categorised landscape expressions under the following headings: sacred structures (temples, monuments or shrines), land patterns (city form or field patterns), cemeteries, the presence and distributions of flora and fauna (cattle in Hindu societies, cultivation of vines) and place names. The invisible evidences of culture, such as ideologies (shared beliefs and ideas), were subject to the same treatment as artefacts (the visible evidences of culture). There was a litany of texts that concentrated on mapping religious or cultural realms and vernacular regions (see de Blij, 1977:148–59, 193–6). Studies of the spatial patterns of religions did move beyond simple static takes on distribution of faiths and also traced the spread and movement of faiths. Indeed, tracing the spread of a cultural attribute or artefact, using diffusion models, was a popular focus within Berkeley School geography (see Price and Lewis, 1993:6). Contemporary geographies of religion, such as Park's (1994) *Sacred Worlds*, devote a significant amount of attention to the spatial shifts of the principal religions, as do much used undergraduate texts (Bergman, 1995:265–83; Fellmann *et al.*, 1995:153–69; Rubenstein and Bacon, 1983:155, 158).

Many of the influential geographers of religion tended to restrict their research to cataloguing landscape influences, yet many of them also expressed a dissatisfaction with this narrow artefactual approach (Deffontaines, 1953:37; Fickeler, 1962:117; Gay, 1971:18; Sopher, 1967:24). Isaac (1959–60:16) made this point well:

> Comprehensive studies of the geography of religion thus far undertaken have failed to be more than simple classifications of types of effects religion has exercised in the landscape, e.g., effect of religion on houses, settlement, agriculture, pastoralism, population movements, etc. . . . While such studies are highly interesting, they fail to bring us nearer to an understanding of the most important problem, namely, in what lies the transforming power of religion upon the landscape and why, in different cultures, has the extent of its effects been so disparate.

These early expressions of discontent demonstrate that geographers were unlikely to be constrained by narrow ideas on what should constitute geographical research and knowledge.

As another example of such an approach of mapping artefacts, we turn to the first geographical analyses of sexuality. Weightman's (1980; 1981) mapping of gay entertainment venues, gay institutions and neighbourhoods corrected the geographical neglect of gays and lesbians, whom Bell (1991:323) referred to as 'the least visible "others" in the eyes and words of geographers'. The early geographies of sexuality made visible the spaces and urban impacts of gays and lesbians. These largely artefactual analyses had the political effect of asserting the existence of gays and lesbians (Weightman, 1981). Later, Winchester and White (1988:41) outlined the way in which the legal and social marginalisation of homosexuals in a homophobic society translated into the spatial marginalisation of gays and lesbians in the cities of the western world. These studies illustrate how artefactual analyses are not necessarily devoid of political insights.

2.2.3 Critique 3: the rustic and exotic

The critique is now made that 'old cultural geography' operated with too limited a definition of what constituted a cultural group. A cultural group was restricted to ethnic groups (linguistic, religious or nationality-based groups) or folk cultures (regional identity and association or affiliations based on locality). Exploring the latter first, Cosgrove (1989:570) asserted that a predilection of 'old cultural geography' was a 'taste for the traditional, the folk' and he referred to a 'muscular disdain for the fey and the metropolitan'. Carney (1990:41–2) noted how music research within cultural geography had retained a folk orientation and very little work was carried out on popular music or music with an urban focus (McLeay, 1998). Work within the folk orientation included discussions of folk regions and of the diffusion of log carpentry techniques, house types, roof structures and fence forms.

The study of ethnic groups tended to be of non-dominant groups. As we discussed earlier, peering at cultural groups different to the western European norm has been a central plank of geography since its foundation. This work assisted with the construction of the idea that these people were 'Others'; people very different from 'Us'. Indeed, the academic focus on the 'extraordinary' aspects of non-Christian, western cultural groups was central to projects of cultural imperialism (Said, 1978). In this context, it is hardly surprising that much of the empirical material of Berkeley School geography had an exotic empirical focus. In Anglo-American cultural and social geography, this included studies of Afro-Americans, Asian–Britons, indigenous Australians, etc. (Bonnett, 1996). This focus on the 'exotic' was also typical of anthropology and sociology at the time. Marcus and Fisher (1986:ix) described it as 'the mindless collection of the exotic'.

Kong (2001) has made the point that there is a need among geographers of religion to return the gaze to the ordinary. With some recent exceptions (e.g. Dwyer, 1999; Naylor and Ryan, 1998), most work in the 'centres' (western geography of Europe and North America) has been lacking in innovation and critical engagement, focused on minority groups (Muslims, Hindus), with few critical attempts to come to grips with large mainline religious groups (often churches) in the USA and UK. For instance, how have churches been taken into other meaning systems, such as through their conversion to commercial space (e.g. tourist sites); what sorts of conflict have arisen in the establishment of churches and secular demands for urban space and how have they been resolved? Critically engaged geographical research on religion has emerged primarily from outside the traditional 'centres'. This includes work about, and emerging from, places such as Australia, Finland, Singapore, Tanzania and Trinidad, perhaps indicative of a commitment to post-colonial analysis (Kong, 2001).

Both anthropologists and cultural geographers now argue for a repatriation of research and a focus on whiteness, on masculinity, on the previously un-problematised ordinaries (Jackson, 1989). 'New' cultural geographers promised to move beyond an analytical emphasis on the vernacular and bizarre. But long before the 'new cultural geographers' began to explore masculinity and whiteness, some of their predecessors had already been 'repatriating' their gaze to ordinary landscapes.

2.3 'New' cultural geographies

A distinction has been drawn between the traditionalist approach of Sauer and others of the Berkeley School and what has become called the 'new cultural geography' (Cosgrove, 1989:566–7; Duncan, 1990:11–24; Jackson, 1989:9–24; Thrift, 1994:109). The most famous exposition of this dichotomy was by the British cultural geographers Denis Cosgrove and Peter Jackson (1987:95):

> If we were to define this 'new' cultural geography it would be contemporary as well as historical (but always contextual and theoretically informed); social as well as spatial (but not confined exclusively to narrowly-defined landscape issues); urban as well as rural; and interested in the contingent nature of culture, in dominant ideologies and in forms of resistance to them.

In this section, we outline the contemporary conception of culture and of cultural landscapes within geography. At the same time, we demonstrate how geographers have been grappling with an understanding of culture as complex and dynamic and how they had come to see the cultural landscape as a process rather than an outcome.

Lester Rowntree warned of the dangers of making a 'simplified dichotomy' between 'old' and 'new' cultural geography. Rowntree (1988:579–82) considered this a generalisation that ignored the theoretical, methodological and empirical breadth of the Berkeley School of cultural geography. Price and Lewis' (1993:10) response to the charge of superorganicism illustrates this concern about generalising the 'old': 'If they [Berkeley geographers] did assimilate the superorganic construct, they did so as individual scholars, not as particles of a reified Berkeley universe.' This is a fair comment and reminds us of the dangers of universalising and generalising. It was ironic that key advocates of 'new' cultural geography who asserted the internal heterogeneity of cultural groups should construct old cultural geography as uniform and fixed. Duncan

(1994:361) later stated that 'good work is done within each [old and new cultural geography], and that none need seek hegemony over the others'.

There are now a handful of concise and balanced summaries of the old-versus-new debates in cultural geography (see Kong, 1997; McDowell, 1994). Linda McDowell (1994:148–50) outlined both the strengths and weaknesses of the work of Carl Sauer and others of the Berkeley School. McDowell (1994:150) pointed to the rich ethnographic and fieldwork tradition, but accepted that Sauer was 'a man of his times' which was manifest in his 'neglect of the wider social, economic and political structures of society'. Lily Kong's (1997) review demonstrates the disciplinary need for such debates. Indeed, this particular debate brought to the foreground the way in which culture was conceived within geography.

2.3.1 The ordinary and the everyday

A decade before the pronouncement of 'new cultural geography', a repatriation of research had begun. In 1979, American geographers Donald Meinig and Peirce Lewis put forward an agenda for cultural geography in which the ordinary landscape and everyday culture became a serious research topic. Lewis' (1979:19–22) focus on common things was an implicit rejection of a prior focus on either the quaint folksy or the extraordinary Other. In similar vein, British geographers Jacquie Burgess and John Gold (Burgess and Gold, 1985) drew attention to popular culture and to the geographical meanings reproduced daily within mass media. These re-orientations can be seen as embryonic forms of 'new cultural geography'.

The emphasis on the everyday or ordinary reflected the research directions within the emerging field referred to as cultural studies. Cultural studies is a cross disciplinary research movement which emerged from Britain in the 1960s and 1970s (see Hall, 1990:11–7). Cultural studies had as its initial empirical focus the ordinary, the banal and the everyday. These were used as entry points to discussions of social relations, exposing relations of domination and cultural oppression. In early and key texts of cultural studies, such as *The Uses of Literacy* (Hoggart, 1957), *Culture and Society* (Williams, 1958) and *The Making of the English Working Class* (Thompson, 1968), everyday cultural forms were championed. Frow and Morris (1993:xxiii) commented that these authors 'saw their task as one of validating the culture of common people' against the canonical values of a British cultural elite. In Chapter 3, especially, we demonstrate how everyday popular cultural landscapes are the outcomes and media of power relations.

Peirce Lewis argued that landscapes can be approached as texts to be read. In this argument, landscapes are likened to a page that has been written on:

> Our human landscape is our unwitting autobiography, reflecting our tastes, our values, our aspirations, and even our fears, in tangible, visible form. . . . All our cultural warts and blemishes are there, and our glories too; but above all, our ordinary day-to-day qualities are exhibited for anybody who wants to find them and knows how to look for them (Lewis, 1979:12).

Every human landscape has cultural meaning, no matter how ordinary it appears. Lewis' statement betrays a legacy of superorganicism through his reference to 'our unwitting autobiography', implying a certain lack of agency. Yet, as we show later, Lewis also referred heavily to the complex and layered nature of cultural meaning within the everyday landscape and this is suggestive of a more contemporary understanding of culture as socially constructed.

Cultural geographers were traditionally of the view that a very useful first step in any research was to find a vantage point from which to look down at the study area. On achieving the necessary altitude, a cultural geographer would then be able to read the landscape, much like reading a book. Sitting atop mountains and 'reading a landscape' did not seem a convincing form of scientific methodology. In the 1970s, during which the social sciences experienced a 'quantitative revolution', the scientific rigour and reliability of this methodology was open to question. Cultural geographers of the time set out to inject some clarity and structure into how landscape interpretations were undertaken by cultural geographers. Donald Meinig (1979a:6) asserted that landscape should be conceived of as a code to be deciphered:

> Every landscape is a code, and its study may be undertaken as a deciphering of meaning, of the cultural and social significance of ordinary but diagnostic features.

Meinig (1979a:6) implied that there were superior and inferior methods of reading the landscape and went on to say that 'anyone can look, but we all need help to see'. Lewis and Meinig set out to develop 'guides to help us read the landscape'. Lewis authoritatively listed seven rules or 'axioms' for those about to read a landscape and suggested the questions that should be asked and how findings should be assessed (see Table 2.1). In the cognate field of landscape architecture, researchers began, in the 1990s, to analyse landscape as text. They searched for the narratives within landscapes and explored broader questions on the grammar of landscape (Hunt, 2000; Potteiger and Purinton, 1998; Spirn, 1998).

2.3.2 Multiple landscape meanings

For Lewis and Meinig, there were multiple meanings within any landscape. Different eras of human occupation left specific traces in the landscape. Layers of meanings are produced within the landscape; these meanings remain in the landscape and they can be peeled back and read (see Box 2.4 and Figure 2.3). In this sense, landscapes are not blank pages onto which culture is inscribed. Almost all landscapes have been substantially impacted on by humans and other biota. Landscapes are better thought of as marked pages onto which new cultural impacts are written. The mark of a previous culture remains, even the attempts to extinguish them leaves traces that can be detected. In describing this, landscape researchers have used the analogy of a palimpsest.

Table 2.1 Rules for reading landscapes

No.	Rule	Explanation	Example
1	Landscape is a clue to culture	Change in the landscape will equate with change in culture and vice versa. Landscape change can lag behind cultural change. A major transformation in the cultural look of the landscape is indicative of a major change in culture	New tastes in cuisine and the entry of new cultures through migration will manifest spatially. Social disharmony and exclusion will manifest as segregation (see Figure 5.4)
2	Cultural landscape equality	No specific landscape feature or cultural manifestation of a group is necessarily more or less important than another, better or worse	The McDonald's Family Restaurant is as instructive to geographers as an artefact of culture; so is the Eiffel Tower
3	Everyday landscape of common things	The ordinary landscape has been little researched. Researchers have been dismissive or disparaging of such common things. If studied with care and without elitism they can tell us a great deal about everyday culture	The rapid spread of McDonald's Family Restaurants (see Figure 2.2) through Australia is reflective of the uptake of American popular culture, as are the everyday shop fronts within malls
4	History and landscape	To read a landscape properly a researcher needs to know something of the history of a place	To interpret graffiti and murals in Belfast properly, one needs to know about sectarianism in Belfast, the Troubles and the historical development of Ireland, especially the north (see Figures 5.3 and 5.5)
5	Geographic context	A landscape or landscape feature can only be understood with reference to the surrounding places and landscapes	A reading of the city of Newcastle (see Figure 2.3) should bring an awareness of nearby coalfields and the nearby presence of a navigable port
6	Physical landscape	The human landscape is related to the biophysical environment. Aspects such as terrain, climate and geology can be important	The river, timber and the presence of coal deposits help explain the location and morphology of Newcastle (see Figure 2.3)
7	Landscape obscurity	The landscape does not speak to us very clearly. Primary messages can be deceiving. Cultural geographers need to ask the right questions and look the right way	Ask the basic questions, especially of ordinary but also exotic landscapes. What do they tell us about how our society works?
8	Landscape silences	The silences of a landscape may be as important as the texts of the landscape. The many silences are integral to cultural processes. They are reflective of power relations and so are clues to our culture and cultural relations	The absence of a mosque in a city suburb may reflect the Islamaphobia that prevented its development. The lack of indigenous culture in a built environment reflects colonial dispossession (see Box 2.4)

Source: Adapted from Lewis, 1979; Sedgwick, 1990:3; 1993:11.

Figure 2.2 McDonald's Family Restaurant near Sydney, Australia. The famous Golden Arch of McDonald's is indicative of the popularity of fast food

Box 2.4   Layers of meaning in the landscape of Newcastle, Australia

The vernacular identity of a place is a composite of the many identities and landscape impacts layered through time. Each era of human occupation leaves temporally specific traces. Dunn *et al.* (1995) identified six layers of meaning within the landscape of the Australian industrial city of Newcastle.

Layer 1   The land around Newcastle was originally associated with the **Awabakal** indigenous people. The Awabakal managed a marine and land economy. They vigorously defended their possession against the British military invasion from the late 1700s and through to the 1820s. Such was the extent of their dispossession and decimation that very little of the Awabakal ways of life remain today in the vernacular identity and landscape of the region. But these silences are themselves indicative of cultural relations, past and present (see Table 2.1).

Layer 2   Between 1804 and the 1820s, Newcastle was transformed into a place of secondary punishment for the British **penal colony** at Sydney. Convicts in the area were set to work extracting coal, timber and lime under deplorable working conditions. Chain gangs laboured on various infrastructure projects and these remain in the landscape. Convict Newcastle was a very male, brutish place, with a social system directly transported from British penal institutions.

Layer 3   With the gradual replacement of convict labour, Newcastle became primarily a place from which coal was extracted and exported. It was a **coal town**. This era was marked by poor labour relations and dangerous working conditions reminiscent of convict times. A strict gender separation of spheres of activity was apparent. Men laboured in the mines and the women toiled within the miners' cottages that surrounded the pitheads. These coal village patterns remain perceptible within the contemporary urban morphology of Newcastle and the city remains a place of coal export.

Layer 4   In 1914 Broken Hill Proprietary Company Limited (BHP) commenced construction of Australia's first integrated steelworks. From this period, the city was not only a mining town, but emerged as a steel and heavy engineering complex. Militant unionism was reinforced by class antagonism and exploitation. Newcastle became an archetypal **industrial city**, with the sights and smells of a 'dirty old town' (see Figure 2.3).

Layer 5   The industrial prosperity and development of Newcastle flagged during the 1970s and especially the 1980s. In April 1997, BHP announced the closure of the steelmaking operations at Newcastle by the end of 1999. This followed a series closures in the heavy engineering areas (e.g. shipbuilding) and in the textiles and clothing industries. Newcastle became a **problem city**. Abandoned industrial sites and demolitions were the landscape manifestations of these changes. The sense of crisis and abandonment and loss of confidence were perceptible in the city's texts (newspaper editorials, cartoons, etc.).

(continued)

(continued)

Figure 2.3 Industrial city, Newcastle, Australia. The landscape can be seen as palimpsest on to which various eras have left specific traces

Layer 6 In the 1990s local authorities began to market Newcastle as a post-industrial city, a **leisure zone**, which was clean and non-polluted and perfect for hosting conferences and conventions. At the same time, NSW state government authorities initiated a foreshore redevelopment. The temporarily obsolete abandoned derelict sites (TOADS) around the harbour were prepared for new investments in housing and commercial activities.

The cultural landscape of Newcastle is an accumulation. The development of a city does not necessarily extinguish the traces of previous eras of occupancy. Clearly, however, the value accorded to a previous era and to its residents will influence the extent to which traces of earlier periods are removed.

Much of 'new' cultural geography uses an interpretative approach that recognises cultural landscapes as interpreted in different ways by people. It is now generally accepted in cultural geography that meanings associated with a cultural landscape are generally multiplicitious (Jackson, 1989:175–7):

> Each person or group views, uses and constructs the same landscape in different ways; these are neither 'right' or 'wrong', but rather are part of the many layers of meaning within one landscape (Winchester, 1992:140).

This perspective draws on important observations by geographers such as Donald Meinig and Jon Goss. Meinig (1979b:33–4) described how each individual will see the same landscape differently (see Box 2.5). Goss (1988)

Box 2.5   Ten types of perception

Donald Meinig (1979b:33–4) made the important assertion that each individual will see the same landscape differently. This variation depends on their looking position or perspective. The term 'polysemy' is used within cultural studies to describe how each individual can interpret film, art and poetry differently (During, 1993:6–7). So it is with regard to a cultural landscape:

> It will soon be apparent that even though we gather together and look in the same direction at the same instant, we will not – we cannot – see the same landscape. . . . Thus we confront the central problem: any landscape is composed not only of what lies before our eyes but what lies within our heads (Meinig, 1979b:34).

To speak of 'ideas in our heads' is to be speaking of ideology. Meinig (1979b:34) listed ten forms of ideology that generated very different readings of a landscape, asserting that 'there are those among us who look out upon that variegated scene and see, first and last':

Nature   the landscape in its pristine or underlying condition
Habitat   reworked nature, fashioned into the home of humankind
Artefact   landscape as bearing the mark of culture
System   landscape as stage for biophysical cycles and social systems
Problem   landscapes needing correction: remediation or investment
Wealth   landscape as a resource or commodity
Ideology   landscape as repository of aspiration, nationalist ideals
History   landscape as the cumulative record of the culture
Place   landscape as locality, sense of home and place
Aesthetic   landscape as subject matter for artistic representation (Meinig, 1979b:34–47).

And, of course, these ten ways of seeing do not exhaust the range of possibilities (Meinig, 1979b:47).

identified four theoretical categories from which buildings have been analysed: as a cultural artefact, as an object of value, as a sign and as a spatial system. In those terms, a mosque could be investigated as an architectural achievement, a piece of real estate, an expression of Islam (or freedom of religious expression) or as the focal point of a Muslim community.

2.3.3 Culture as a way of life: landscape as a process

'New' cultural geographers were critical of earlier conceptions of culture through which people were categorised into a limited number of ethnicities or other categories, much in the way that animals and plants are allocated by taxonomists into certain classes, orders and species (Duncan, 1980). However, ways of human life can be seen as dynamic. Human beings are able to change their culture or way of life. Contrary to the superorganic conceptualisation, cultures are not static. Cultural studies researchers suggest that they continually evolve and they are contested (Clifford, 1986:18–19). If cultures are dynamic, then any attempt to sort people by their way of life will always be outdated by the rapid changes in individual cultural affiliations and the constant alteration of cultures.

Cultures, or ways of life, comprise two important components. Cultural groups are marked by shared performances (behaviours/rules) and shared ideology (beliefs and understandings). These shared performances and ideologies are often provisional and are internally contested, but they are nonetheless the basis of group affiliation and cultural identity. Cultures develop constantly in concert with the everyday experiences of individuals. For example, Pulvirenti (1997) explained how Italian–Australians are very different cultural subjects than are Italians in Italy. This is because of migration and settlement experiences and exposure to differing sets of societal influences, so much so that:

> Culture cannot be thought of as that brown paper parcel transported with migration, and preserved. . . . Experiences of migration, which extend well before the journey and after it, produce and reproduce culture continually. . . . This culture does not depend on an understanding of identity as 'essence', but on subjectivity as it is created through these experiences (Pulvirenti, 1997:38).

Cultural subjects, or individuals, are constantly changing. New ways of life, or what we call cultures, are constantly being made and remade.

Stratford (1999:5) observed that if cultures are dynamic, then it is appropriate to refer to the cultural landscape as a process. Everyday landscape features are used to (re)construct culture and identity. For example, in post-colonial Singapore, street names were renamed to exorcise colonialism (Chapter 4). Despite significant global criticism, the Taliban government of Afghanistan destroyed ancient Buddhist statues in early 2001. They did this in an attempt to erase the presence of non-Muslim belief. Anderson's (1987; 1993) work on Chinatowns and the constructions of Chinese-ness in North America and Australia demonstrated the way in which identities are manifest within landscapes. Identities are also reinforced through the landscape, in the media and other cultural products (Duncan and Duncan, 1988:124; McDowell, 1994:163). Landscapes are important to the (re)construction of identity.

2.3.4 Social construction theory

The 'social construction approach', or what has also been referred to as the 'constructivist' approach, has been argued for forcefully within cultural geography (Jackson and Penrose, 1993; Kobayashi and Peake, 1994; Watts, 1991:8–9). Lily Kong advocated further application and development of social construction theory within cultural geography:

> What is needed is, of course, continued efforts at theorisation, but also much more empirical work that takes on board reconceptualisations of culture as an active and negotiated construct and centralises social construction theory (Kong, 1997:183).

Advocates of social construction theory argue it is a conceptual framework that readily embraces the dynamic nature of culture.

The basic premise of a social construction approach is that categorisations of humanity – such as notions of race, ethnicity and gender – are outcomes of human thought and action. They are social constructions and not natural or primordial givens. These categories did not evolve out of the ether with humans; rather they were constructed by us. Cultural identity is, then, socially constructed. In their edited collection *The City in Cultural Context*, Agnew *et al.* (1984:1) stated that most cultural geographers now accept that 'culture is created by thought and actions of both historical and living populations'. Clearly, the conception of culture as socially constructed has been accepted since at least the mid-1980s by cultural geographers.

Conceiving of culture and identity as constructed is not to say they are not important or unreal. Indeed, recent world conflicts demonstrate that social constructions of 'race', religion and nation have substantial impacts on the lives of many people. Social constructions underpin group affiliation and mobilisation, including exclusions and atrocities (see Chapters 4 and 6).

Having accepted that identity is a social construction, the social construction approach demands that we deconstruct identities (Kobayashi and Peake, 1994:230). This involves unsettling extant social constructions, whether they be 'race' or gender based, for example. It should involve an exposing of dominant ideologies that underlie such constructions, the institutions which aid those constructions and the groups who are privileged by them.

Dominant groups often seem able to create structural oppositions in which they conceive of themselves as 'normal' and ordinary while subordinate groups are treated as 'Others' or extraordinary (Chapter 4). Dominant groups assume that their own identity is unproblematic:

> Dominant groups still manage to efface their own actions by implying that they are somehow outside the process of definition (Jackson and Penrose, 1993:18).

An example of this would be media references to 'Australians' and 'ethnics' as opposites. These depictions construct a border, with Anglo-Celts or whites as the 'Self' and members of all other ethnic groups as 'Other' (see Goodall *et al.*, 1994:54) (Chapter 5). A social construction approach would disrupt these assumptions of 'Self' and 'Other'.

Similarly, the identities of places are not givens. They are also socially constructed outcomes (Jackson and Penrose, 1993). Doreen Massey (1992:11) noted that attempts are constantly made to fix the identities of places, for instance, through place marketing or expressions of nationalism. However, places are complex sites of meaning:

> A place is formed in part out of the particular set of social relations which interact at a particular location. And the singularity of any individual place is formed in part out of the specificity of the interaction which occur at that location (Massey, 1992:11).

This reveals the obvious but previously unacknowledged nature of place identity as dynamic, provisional and contingent (Massey, 1992:12–4). The unfixed and dynamic nature of place identity is analogous to the dynamic nature of the cultural landscape.

2.4 Texts, expert decoders and real landscapes

The geographical focus on texts about landscapes and the treatment of landscape as if it were a text have been strongly associated with 'new' cultural geography and with post-structuralist theory. Both 'new' cultural geography and post-structuralists have been heavily criticised for this textualism. Reviewers and guest editors have been critical of such a textualist approach within geography, asserting a lack of empirics and political weakness (Badcock, 1996; Gregson, 1993; Thrift, 1994). Nicky Gregson (1993:529) was depressed by a geography that was 'seduced by text' and that had 'lost the desire to say anything about this empirical social world'. Nigel Thrift (1994:110) was concerned that the textualist approach 'does not give sufficient room to issues of power and oppression' and that there is often 'little sense of a world out there'. Martin Price and Marie Lewis criticised the 'new' cultural geographers for their 'descent into a self-contained theoretical universe' and for a debilitating dependence on textual approaches. They argued that 'Duncan's insistence on viewing all landscapes as 'texts' comes dangerously close to demanding the intervention of an 'expert decoder' – the new cultural geographer' (Price and Lewis, 1993:12).

Perhaps the closest evidence of such dallying with textual analyses may be found in Duncan's (1990) *The City as Text*. While in general, Duncan does an illuminating job with bringing together the different symbolic readings, his analysis of the symbolic content of Kandyan landscapes is sometimes burdened by his concern with drawing linguistic analogies: the identification of tropes in the form of synecdoches, metonyms, similes, allegories and so forth. For example, when he suggests that the lake in Kandy is a synecdoche for the Ocean of Milk in Hindu thought, the linguistic analogy reveals little of what the lake actually symbolises. It is only when he explores the symbolic meanings without the linguistic baggage that the parallels become clear. In fact, it seems possible to achieve the goal of uncovering ideological content without the semiological superstructure. In general, the risk is great of getting caught up with identifying the landscape and buildings too literally as language and texts and can result in attempts to establish the grammar of space, while forgetting what should be the main concern: uncovering the ideological meanings of landscapes.

Despite the cautionary notes, it is not at all clear that geographers can rise above textualism. Luce Walton (1995:62) has explained that the only external or real world accessible to us is that which we construct through our interpretations:

> Accepting landscape as 'text' (in the contemporary sense of the term) implies that what we can know is not a given, universal, 'authentic' world, but an epistemologically mediated reality, constructed linguistically as well as materially.

Put simply, any description of a landscape or cultural group is text-like (Walton, 1995:62, 64; 1996:100). The post-structuralist theorist Jean-François Lyotard (1984:15, 40) similarly argued that, while there was a social reality beyond text, social reality was actually orchestrated and lived through text. However, this should not mean that there is no such thing as a real world independent of the cultural geographer, but only that we have no access to it as a pre-interpreted reality (Hastings, 1999:7; Jacobs, 1999:203; Walton, 1995:62, 64; 1996:99).

Cultural geographical analysis of texts need not be entirely divorced from the 'real world out there'. Discursive representations of landscapes often can have real material consequences and it calls for the skills of a cultural geographer willing to engage with textual representations as well as with 'real landscapes' to make sense of cultural phenomena (Dunn, 1997). Thus, 'new' cultural geographers are all the more effective if they pay attention not only to texts and landscapes, but also analyse how such texts are consumed and with what impacts (Burgess, 1990; Dunn and Winchester, 1999:179–80, 188–90; Goss, 1988:398; Kong, 1997:181). To understand the cultural climate of 1970s' Singapore, for example, a textualist approach would throw up significant insights into the discursive reality in which pop and rock music resided. The state, aided by agencies such as the mass media, constructed pop and rock music as a 'folk devil', attempting to whip up a 'moral panic' (Cohen, 1973). Media coverage and commentaries, public speeches, parliamentary debates and other such discursive realms were participatory to such constructions and lend themselves to detailed textual analyses. But such textual representations are not divorced from the 'real' moral geographies on the ground and, indeed, have real material consequences. The inscription of moral geographies at the large scale – that of the national and global – is evident in the demarcation of national boundaries as the boundaries within which morality resides and beyond which belong negative decadent forces. Such ideological constructions resulted in impacts on who and what were allowed into the country. At the scale of the local, moral geographies were constructed: first, in terms of places and associated activities which are thought to be morally damaging (nightclubs and discos) and, second, at the spatial scale of the individual self, where the body becomes the site of moral judgement (long hair = immoral) (see Chapter 7). Again, such ideological constructions purveyed in the discursive and textual realms had significant material consequences in the closure of clubs and the banning of long hair.

These debates within geography about textualism mirrored debates between post-structuralist and structuralist theorists. The Marxist structuralist Frederick Jameson (1984:xviii) criticised post-structuralist practice for a 'falling away of the protopolitical vocation' and characterised its replacement as 'a commitment to surface and to the *superficial* in all senses of the word'. Similarly, the Marxist geographer David Harvey (1992:315–6; 1996:438) has asked questions about the political efficacy of post-structuralism, stressing the disadvantages of a textualist approach and concluding that 'practical politics begins' where 'discursive reflection ends'.

There is little doubt that research must be grounded in a concern for the outcomes of the ideologies that inform and constrain the spatial expression of culture. The material outcomes of ideologies must be of concern to cultural geographers and these outcomes should be guides in evaluating the desirability of such ideologies and for whom. For instance, if dominant ideology militates against the spatial activity or expression of a subordinate cultural group, then the ideology is challengeable from the perspective of the latter group. This is, of course, not to say that texts are not important, but rather that they are made important by the crucial role they have in (re)producing power relations.

The post-structuralist influences within much of 'new' cultural geography emphasise individual interpretations of texts and landscapes. Despite the individuality of interpretation, there are still common or shared interpretations and meanings. By borrowing Stock's (1986:294) idea of 'textual communities', Duncan and Duncan (1988:120) pointed out that there are groups of people who share understandings of landscapes:

> Stock . . . tempers the radical individualism of poststructural theory by positing the notion of a textual community. He defines a textual community as 'a group of people who have a common understanding of a text, spoken or read, and who organize aspects of their lives as the playing out of a script'.

Research on textual communities attempts to uncover the ideologies and institutions that lie behind common interpretations: what Stock (1986:297) referred to as the 'wider hermeneutic network'. For example, the widespread opposition to Islamic mosques throughout Sydney in the 1980s and 1990s demonstrated that people were sharing a reading of Islam and also of what was an acceptable 'Australian' (Chapter 5). Newspapers, radio and television images in Australia have had a decisive role in the construction of Islam and have been central to the mediation of negative stereotypes that underpin the opposition to mosque development. Stereotypical portrayals of religious groups such as Muslims, which circulate in the international and national media, feed into the locality, the street and the home (Dunn, 2001). Shared interpretations can lead to organised and common resistances to cultural spatial expressions.

2.5 Approaching the cultural landscape

Clearly, studies of cultural distribution and diffusion, and other approaches that could be described as artefactual, are legitimate methods for cultural geographers. However, such work must be cognisant of the power relations that permeate society (Chapters 4 and 5). At the turn of the new millennium, cultural geographers are aware of the dynamic nature of culture and of the contested and fluid nature of meanings that are derived from landscapes and cultural products or texts. The cultural landscape is a complex process that has an important role in the re-production of everyday culture (Chapter 3). The everyday cultural landscape, and the texts of those landscapes, are the stuff of cultural geography. As we show in the following chapter, they are revelatory and constituent of our everyday ways of life.

Chapter 3

Landscapes of everyday popular cultures

In this chapter, we begin the exploration of landscapes by focusing on that which is familiar and close to us, namely, the landscapes of our everyday popular cultures. Many of the themes that we discussed in the previous two chapters are exemplified here: landscapes as expressions of culture; and landscapes as representations which construct and reinforce identities, but also make possible the contestation of identities and meanings, thus rendering landscapes the sites of resistances to repressions.

While cultural geographers influenced by the Berkeley School tended to focus their gaze on the exotic and the antiquarian, in recent years increasing research attention has turned to landscapes of everyday life. These take the form of ordinary landscapes in our daily routines, such as street names, shopping centres and neighbourhood spaces, which we examine in later chapters, as well as landscapes of popular cultures, such as those of food, film and music, which we focus on in this chapter. This focus on landscapes of everyday popular cultures returns the gaze to the ordinary and commonplace. The landscapes that we live in and interact with on a daily basis are not to be understood as 'innocent' and value free. They are as deeply implicated in the maintenance and challenging of ideological meanings and power relations as landscapes of the exotic and spectacular. In so doing, we retrieve from academic, and particularly, geographic, oblivion that which is popular and thus looked on with some disdain because it is deemed 'not serious' and 'ephemeral'.

3.1 Defining the popular

Before exploring the landscapes of everyday popular culture, we take readers first through the multiple conceptions of popular culture. The *Journal of Popular Culture*, as an academic 'gatekeeper' moderating conceptions of 'popular culture' via the papers it accepts, approaches popular culture as products designed for mass consumption, although such a conception may be faulted for its sheer inclusiveness. Other inclusive conceptions suggest that popular culture is constituted by whatever people do in their leisure time. Yet others suggest that popular culture refers to those cultural forms that spring directly from the concerns and consciousness of the people among whom it is circulated. This, however, becomes potentially problematic. What are conventionally regarded as 'elite' rather than popular culture (such as classical music and opera) could fit this conception. To narrow this, reference might be made to the everyday practices, experiences and beliefs of the 'common people', that is, those people who do not occupy positions of wealth and power, the non-elite, whether defined in economic or social terms (Burgess and Gold, 1985:2; Hall, 1981:238). In this sense, 'popular culture' might be conceived as that which 'belongs to the people' and that which is 'widely favoured' and 'well liked', such as popular music.

We scope the chapter by focusing on popular culture as that which appeals to 'common people' and is associated with their everyday practices, experiences and beliefs. The landscapes of everyday popular cultures are therefore the landscapes associated with everyday popular practices and of common people. We use as case studies various aspects of quotidian practice, namely, foodscapes, streetscapes, musicscapes and filmscapes. The first two afford opportunity for us to examine 'real' landscapes while the last two provide access to analysis of 'represented' landscapes. Where appropriate, we will surface the intersections of 'real' and 'represented' landscapes. We will also engage with intersecting scales, illustrating yet another theme in this book. We organise our case study material, using as starting points, the international, national and local. Within each section, however, we highlight how phenomena that appear to embody inherently and primarily the workings at one scale become inexorably drawn into the goings on at another scale.

3.2 Everyday landscapes of cultural globalisation

The world has become ever more tightly enmeshed in a single global economy, intertwined by vast networks of telecommunications, and encircled by the rapid movement of goods and the transnational migrations of populations. Video and audio cassettes and satellite dishes bring information with great rapidity into an oasis village in the Sahara, a neighbourhood of artisans in Nicaragua, an exclusive boarding school in New Hampshire, an immigrant pub in Brixton, and an executive suite in Tokyo (Schiller, 1994:1).

For those who argue that the growing global reach of capital, media, technology, people and ideology (Appadurai, 1990) has caused the construction of a single global economy and a homogenised cultural existence characterised by homogenised commodity consumption, the power of globalising forces is very real. In this 'McWorld', we eat the same food, engage in the same sports and sporting events, listen to the same music and watch the same movies. In such a conception, the landscapes of everyday popular cultures are the same the world over.

Much of these arguments about the homogenising effects of global forces equate globalisation with Americanisation (Box 3.1). They echo the view that a process of cultural imperialism is at work in which the cultural spaces of (Third World) nations are being broken down by the 'economic and political domination of the United States . . . thrust[ing] its hegemonic culture into all parts of the world' (Featherstone, 1993:170). This is viewed as deleterious for local cultures because of the assumption that all local cultures would eventually

Box 3.1   The McDonaldisation of the world

The spread of well-known brands (Coca-Cola, Levi, McDonald's, Pepsi-Cola, Starbucks Coffee) has been perceived, and sometimes analysed, as indicators of the spread of American culture and also of a certain form of consumer capitalism (Weightman, 1988:59, 64). For example, McDonald's restaurants have come to 'symbolize the "American Way"' (Fishwick, 1995:13).

Laurence Carstensen (1995) used diffusion models to chart the expansion of McDonald's Family Restaurants in the United States. In 1955, there were two principal innovation sites of the modern McDonald's (Fishwick, 1995:13–14; McDonald's Corporation, 2001a:7–8), one in Illinois (Des Plaines) and the other in California (San Bernadino). In the latter part of the 1960s, there was a tremendous expansion of McDonald's restaurants within the United States, with 150 new restaurants built in the state of California in a four-year period. In 1977 there were 5185 McDonald's restaurants across the globe (Carstensen, 1995:121). While these restaurants were mostly in the United States, the corporation had 394 outside of the United States and was building another 45 in international markets (Carstensen, 1995:121; Fishwick, 1995:13–14). From this period on, the expansion of McDonald's was very much a global phenomenon.

The expansion of McDonald's principally followed a hierarchical expansion diffusion. Cartensen's (1995:124–6) research revealed that larger places in the USA developed McDonald's restaurants before smaller settlements. The trend was for McDonald's restaurants to be located first in larger cities and then also in key stops along major roadways, followed by much smaller localities.

The first McDonald's Family Restaurant in Australia was opened in 1971 in the western Sydney suburb of Yagoona. By the late 1970s the rate of McDonald's growth in Australia was quite rapid. In 1978 alone 23 McDonald's Family Restaurants were opened in Australia (see Figure 2.2). That rate of expansion has been quite constant. Twenty restaurants were opened in the 2000–2001 financial year (McDonald's Corporation, 2001b). There are now 704 restaurants throughout Australia.

The McDonald's Family Restaurant has spread to almost every corner of the world. In mid-2001, there were 28,203 McDonald's restaurants spread across 120 countries (McDonald's Corporation, 2001b). Often this has been assisted by the thawing of international politics, with the end of the Cold War and the rise of *glasnost* (Fishwick, 1995:21).

McDonald's opened restaurants in Moscow in 1990 and Beijing in 1992. On the first day of operation, 30,000 Muscovites queued up, as did 40,000 Chinese (McDonald's Corporation, 2001a:12). McDonald's has been quite ebullient about its entry into former Iron Curtain countries (McDonald's Corporation, 2001a:12–3). In recent years, the greatest expansion has been in the Asian and Pacific regions. In the financial year of 2000–2001, 200 of the 593 new McDonald's restaurants were established in Asia–Pacific (McDonald's Corporation, 2001a).

give way under the relentless modernising force of American cultural imperialism. In the realm of music, for example, it is assumed that local musical styles will be effaced. In the context of food, it is believed that local cuisines will be overtaken by American fast food. Yet, of course, there is more than one prominent centre of popular culture production. Think of Egyptian cultural domination in North Africa or Brazilian culture in South America, for example (Appadurai, 1990:295).

Regardless of the specific cultural imperialist, the argument that globalisation leads to the invasion of certain cultural spaces is further developed in Meyrowitz's (1985) contention that with improving telecommunications, greater flows of information and images, the sense of collective memory and tradition of localities will be obliterated, such that there is 'no sense of place'. As places become increasingly like one another, the importance of locality is diminished. Followed to its logical conclusion, this cultural homogenisation of 'here and there' implies that music, food, films, television and other cultural forms lose their character generated by a locality of origin.

Take a specific instance of such an argument. It has been posited that music, in particular rock music, has been subject to global processes of commercialism, characterised by commodification (Adorno, 1992). This has led to a process of cultural homogenisation (Cohen, 1991:342), associated variously with conditions of placelessness and timelessness (Meyrowitz, 1985; Wallis and Malm, 1984). Such global processes are said to have caused music to lose its aesthetic value (Adorno, 1992; Smith, 1990:177).

Apart from the cultural homogenisation that is said to come with the spread of western rock music, a second example to illustrate cultural globalisation is the expanded presence across the globe of McDonald's restaurants. If landscape is a clue to culture (Chapter 2), then what does the spread of McDonald's restaurants tell us about culture? Is it producing a world of cultural sameness or do they create diversity, spawn cultural innovation and feed dynamism? For those who have studied the 'McDonaldisation' of the world, the spread of McDonald's has been read as an indicator of prevailing economic and political hegemony of the last superpower – the United States (Ritzer, 1992). Clearly, many people are concerned about such diffusion of tastes and cuisine (Ritzer, 1992). However, many more people have a preference for their products and express this through their patronage. McDonald's serves more than 38 million customers each day (McDonald's Corporation, 2001a:13). This global diffusion is reflected in the clear corporate strategy of other such companies, such as Starbucks Coffee. A Starbucks executive, for example, stated that:

> We see ourselves as the premier purveyor of the finest coffees in the world. And, frankly, we'd like to see a Starbucks coffee shop in every neighbourhood where people like to enjoy great coffee (quoted in Smith, 1996:502).

That cultural globalisation leads to a homogenisation of everyday cultural landscapes is, however, a contested idea. Returning to the examples of both music and food, we illustrate local resistances that suggest cultural imperialism is not an unmediated and one-dimensional process.

In the context of music, some argue that even the most ostensibly globalised elements of music making are subject to distinctive national and local influences (Malm and Wallis, 1993). Indeed, some suggest that imported pop can be a resource of new sounds, instruments and ideas which local musicians use in their own ways to make sense of their own circumstances (Frith, 1989; Hatch, 1989). In this sense, the position that Anglo-American commercial music is destructive to indigenous music is a fundamentally untenable one. The process at work is actually one of transculturation, a 'two-way process that both dilutes and streamlines culture, but also provides new opportunities for cultural enrichment' (Wallis and Malm, 1987:128). Transculturation is apparent when musicians are influenced dually by their own local cultural traditions and by the music industry's transnational standards. The result is local music with a transnational flavour or transnational music with a local flavour (Wallis and Malm, 1987:132). The important question, then, is how far the transculturation process will go and what implications this has for the production of culturally differentiated music that is at the same time international in nature (Wallis and Malm, 1987:133). Some have argued that transculturation in fact gives rise to 'third cultures' which draw on the culture of the parent country but also take into account local cultures and local practices (e.g. Featherstone, 1993).

Dick Lee, a Singaporean singer, composer and songwriter, illustrates this transculturation through his music, a fusion of east and west, local and global. Lee's music reveals how he has been influenced by his own local cultural traditions as well as the music industry's transnational standards. While some elements of east and west, local and global are lost in the process, others are retained or imbibed. Indeed, Singapore's musicscape as shaped by Lee is one in which a new sound is created, sometimes termed 'Singapop'. In this creation of a new sound, identified as distinctively Singaporean, tradition is (re)invented.

What are the evidences of such transculturation and (re)invention of tradition? Instrumentation, styles, language, costumes and guest artistes have all been called into play. In his 1984 album, 'Life in the Lion City', Lee mixed traditional instruments with synthesisers and claimed this to be his first introduction to the search for a new identity via music. This has become a recognisable feature of his music, described as blending pop funk beats with Asian instruments and a fusion of different idioms: 'a bit of pop, a bit of fusion, a little jazz and a subtle working in of the ethnic element' (Kong, 1996a). In addition, Lee's music has often been described as 'anglicised' versions of local songs. For example, a Chinese folk song, 'Little White Boat', is given English lyrics; a traditional Malay folk song, 'Rasa Sayang', is infused with a modern rap; and traditional folk tunes (e.g. 'Alishan', 'Springtime' and 'Cockatoo') and English pop styles are cleverly incorporated. As one critic suggested, Lee's 1989 album 'The Mad Chinaman' is a hybrid of traditional folk melodies and modernised techno-rhythms (Kong, 1996a).

Lee's mixing of cultures is closely reflected in language and Singlish (Singapore English) is an excellent example of how a colonial and international language has been adapted to local use and Chinese and Malay elements have crept into usage to result in a colloquial form of English. Lee uses Singlish abundantly in his lyrics and his litany of Singlish expressions ('shiok', 'lecheh', 'lah', 'leh', 'meh' and others) amid otherwise standard English attest to the working of transculturation. This has evolved from Lee's earlier technique of singing in a mix of languages within the same song, such as tagging a Chinese song at the end of an English pop tune (Kong, 1996a).

Further evidence of transculturation is seen in Lee's stage presentation and choice of guest artistes. At a local discotheque, performing his tracks from the 1991 album 'Orientalism', Lee is dressed à la Elvis Presley, with sideburns and all (*The Straits Times*, 24 August 1991). This display of western influence while singing about the Orient is a seeming paradox, but a reconcilable one when one considers that Lee is concerned with the fusion of east and west, local and global, tradition and modernity. Further evidence can be found in one of Lee's earlier efforts to encapsulate Singapore in song, the 'Life in the Lion City' album referred to earlier. The rap in the song of the same name about local life is performed by American friend Melise. Some may argue that this undermines the local message, but it may well be said that it reveals the inevitable transculturation process (Kong, 1996a).

Such transculturation is also evident in other cultural forms. Returning to the example of McDonald's, to the casual observer, a McDonald's in one cultural landscape looks much the same as those in very different cultural contexts. This brand, alongside others, seems able to transgress cultural difference or pay little heed to local cultural contexts. Nonetheless, some aspects of cultural adaptation are perceptible. McDonald's restaurants in Arab countries have incorporated halal-only ingredients, especially for their beef patties, and in Israel the food is 'kosher'. The Big Mac in India is made from lamb and is called the Maharaja Mac (McDonald's Corporation, 2001a:13). These are attempts to overcome cultural circumstances that would curtail the consumption of these products. McDonald's Corporation has also adopted a policy of incorporating local cultural features into the design of restaurants, often through adaptive reuse of existing buildings. However, as Fishwick (1995:20) noted, everything behind the counter is globally uniform.

The experience of McDonald's extends beyond this responsive transcultural intersection of the global and local. In other contexts, the intersection takes on a much more critical and sometimes confrontational turn. In Australia, resident action groups have protested the ecological practices of McDonald's. In the Australian city of Newcastle, the city council responded to such criticism and insisted that McDonald's introduce more environmentally sound serving and packaging to their inner-city facility. McDonald's resisted this imposition, but were unable to have all their expansion plans approved.

Protesters in the Blue Mountains, to the west of Sydney, successfully campaigned against the establishment of a McDonald's restaurant. A protest leader, representing local restaurants, cafés and other food outlets, made the following comment:

The Blue Mountains . . . is seen as a haven, a place to appreciate nature and enjoy the environment. . . . Do we want the Upper Mountains to first have a McDonald's then a KFC, a Pizza Hut, a Red Rooster strung along the highway? If I were a visitor I would say 'why did I bother coming up here, this is just what I was trying to escape from' (*The Blue Mountains Gazette*, 18 September 1996:5).

According to this opponent, McDonald's was antithetical to environmental appreciation. The protesters were using a particular construction of their own locality – a landscape of environmental beauty – as an argument to oppose the establishment of a future competitor.

Another example of such concern and resistance targeted at the way in which 'international brands' operate without paying attention or being sensitive to the local circumstances in which the product is consumed, thus squeezing out local products, local icons and local jobs, is raised in Bolivia over McDonald's. Concern was raised in Bolivia that McDonald's was not using Bolivian potato for the french fries (Moriarty, 1999:21). While the burgers used Bolivian beef, it was said that the company was compelled to use Canadian potatoes in order to ensure global uniformity in the french fries (Moriarty, 1999:21). Critics have also turned on those people who choose to consume McDonald's french fries. Moriarty (1999:21), for example, lamented the popularity of McDonald's in Bolivia:

I guess my real problem with McDonald's here is more cultural than it is environmental or nutritional. Too many Bolivians believe that all things foreign are inherently superior. The arrival of McDonald's last year gave upper-class Bolivians a unique opportunity to live out that belief.

Yet, Moriarty went on to admit that he very much liked McDonald's comfort food. It was 'familiar, quick and fairly unlikely to give me parasites'. The elite of La Paz and expatriates would crowd around the McDonald's and Domino's Pizza establishments and dream they were in America. Goldberg (2000) went so far as to assert that the arrival of the Golden Arches is a sign that a nation has raised its standards and achieved economic maturity. McDonald's restaurants emphasise sanitary conditions, low prices and ordered service (McDonald's Corporation, 2001a:7, 15). In a highly ethnocentric argument, Goldberg noted how McDonald's in Beijing had brought sanitary conditions unknown in 'most indigenous Chinese eateries' and that customers were taught to wait their turn in ordered queues. Having a McDonald's restaurant is perceived by some receiving countries to be a mark of having become a real country:

A real country, like America – where there is one McDonald's franchise for every 30,000 people (the exact ratio the Founders intended for congressional representation). Around the world, the company has more than 25,000 outlets in 119 countries. Clearly McDonald's is giving people something they want (Goldberg, 2000:32).

In sum, the McDonaldisation of the world is a contested process and there is no singular embracing of or resistance to it. Just as some perceive the presence of McDonald's to signal that a country has arrived, McDonald's restaurants have also been favoured targets of some protesters to world economic forums and the meetings of the World Trade Organisation (WTO). A right-wing American commentator deplored this critique of McDonald's in the following fashion:

> In fact, if you're anti-WTO, anti-America, anti-capitalism, anti-globalisation, anti-biotech – pretty much anti-anything – its likely that you view Ronald McDonald as your sworn enemy and that your most persuasive public arguments involve taking a bad word and putting 'Mc' in front of it. Signs denouncing 'McGreed', 'McPollution', and corporate 'McDomination' are staples of left-wing protests everywhere (Goldberg, 2000:29).

Indeed, in this sense, McDonald's, more than any other company, has become synonymous with cultural globalisation and with the threat of global cultural uniformity.

Whether it is musicscapes or foodscapes, cultural globalisation has not entirely resulted in complete cultural homogenisation. Instead, local forces are often reasserted and/or a process of transculturation and hybridisation takes place.

3.3 Everyday landscapes of national identities

Moving from the global scale, we turn in this section to a discussion of the 'national'. The literature on national identities acknowledges that such identities are often, if not always, deliberate constructions rather than 'natural' givens. The political discourses that seek to define hegemonic visions of 'nation' and 'national identity' have been explored in a growing literature by geographers, among others. Daniels (1993), in *Fields of Vision*, illustrates how landscapes in various media have articulated national identities in England and the United States from the later eighteenth century. Similarly, Kong and Yeoh (2002) illustrate how landscapes in Singapore are an integral part of an explicit and deliberate construction and deconstruction of national identity, achieved via a variety of strategies, from political rhetoric to specific government policy to cultural (re)production. Such constructions are, of course, also debated, criticised and resisted, illustrating how hegemony is never total (see Chapter 4).

In this section, we situate our discussion of landscapes of everyday popular cultures within a broader discourse on the constructed nature of 'national identity' and illustrate how landscapes in various popular cultural forms (as well as popular culture in particular landscapes), constitute socially constructed versions/visions of nations. We use the examples of filmscapes, musicscapes and foodscapes to make the case.

3.3.1 Filmscapes

A handful of geographers have commented on the place-making power of film (Aitken and Zonn, 1993; Benton, 1995; Dunn and Winchester, 1999;

Gold, 1985). Feature films construct potent images of places and landscapes, including both urban and 'natural' landscapes (Kennedy and Luckinbeal, 1997:42–5). Because of the powerful role of film in the cultural construction of place and landscape, they are a site of political contest. Debates revolve around the representativeness, authenticity and cultural motivations of film.

Films are a form of text or cultural product that are recognised as having particular influence in the social construction of national identities. National governments dedicate considerable resources to support their indigenous film industries. It is considered important that citizens consume films that reflect local cultural landscapes:

> Australia's own cultural self-understanding will be crucially influenced by the way in which our artistic products portray our environment and social context. If we want to protect local mores, idioms, symbols and values, then these must be reflected in our cultural products. The portrayal of varied aspects of one's own traditions as legitimate, valuable and entertaining is important in its own right (Abbey and Crawford, 1987:152).

In 1997, 29 Australian-made films were released, which was well above the average (Australian Film Commission (AFC), 1998:76, 121). However, of the 282 films screened in that same period, 185 films were made in the USA (AFC, 1998:147). The imbalance of consumption is more stark than for distribution. In 1997, only five per cent of national box office takings were from Australian-made films. The best ever box office shares were in 1986 (24%) and 1988 (18%), the years when the iconic films *Crocodile Dundee* and its sequel were released. It has been estimated that as much as 85 per cent of on-screen time in Australian cinemas is dedicated to US-made films (see Crofts, 1989:130).

There is a long-held assumption of vulnerability to US cultural dominance in Australia (O'Regan, 1988:167). Australian governments have also invested about A$100 million annually in film production in the 1990s (AFC, 1998:20). Both conditions have led to an expectation that film products should contribute to the construction of national identity. The manner in which the Australian landscape and people are represented in film is therefore keenly assessed by commentators and opinion leaders. This influences the films produced and the constructions offered within them.

While there is some consensus that it is culturally important to have a sustainable film industry, there is much less agreement on what the appropriate constructions of national identity should be. Should Australian films profess images of white settlers battling the bush or should films be set in the urban realm where the vast majority of Australians live? Defining 'Australianness' is a fraught political process (Rowse and Moran, 1984:229–33, 241). Researchers in Australia have nonetheless identified a dominant stream of film representations. A canon of period pieces has evolved, involving whites (mostly men) in bush landscapes or at war (Rowse and Moran, 1984:237). This includes *The Man from Snowy River*, *Gallipoli*, *Breaker Morant* and *Picnic at Hanging Rock* (see Table 3.1). Rowse and Moran (1984:241) commented

Table 3.1   Top 30 Australian-made films, by gross box office in Australia, 1966–June 1998

Rank	Title	Release date	Box office ($)*	Setting(s)
1	*Crocodile Dundee*	1986	47 707 045	Wilderness and New York
2	*Babe*	1995	36 776 544	Rural
3	*Crocodile Dundee II*	1988	24 916 805	Wilderness and New York
4	*Strictly Ballroom*	1992	21 760 400	Suburbia
5	*The Man from Snowy River*	1982	17 228 160	Rural/bush
6	*The Adventures of Priscilla, Queen of the Desert*	1994	16 459 245	City, small town, desert
7	*Muriel's Wedding*	1994	15 765 571	Suburbia
8	*Young Einstein*	1988	13 383 377	Rural
9	*Gallipoli*	1981	11 740 000	War
10	*The Piano*	1993	11 240 484	Rural
11	*Mad Max II*	1981	10 813 000	Desert
12	*Green Card*	1991	10 585 960	City (USA)
13	*The Castle*	1997	10 326 428	Suburbia
14	*Shine*	1996	10 167 416	City
15	*Pharlap*	1983	9 258 884	Rural and city
16	*The Man from Snowy River II*	1988	7 415 000	Rural
17	*Lightning Jack*	1994	6 439 819	Rural
18	*Mad Max*	1979	5 321 000	City
19	*Reckless Kelly*	1993	5 444 534	Rural
20	*Picnic at Hanging Rock*	1975	5 134 300	Rural
21	*Breaker Morant*	1980	4 735 000	War
22	*Alvin Purple*	1973	4 720 000	City
23	*Mad Max, Beyond the Thunderdome*	1985	4 272 802	Desert
24	*Puberty Blues*	1981	3 918 000	Suburbia
25	*Malcolm*	1986	3 483 139	City
26	*The Delinquents*	1989	3 370 650	City
27	*The Sum of Us*	1994	3 327 456	City
28	*Romper Stomper*	1992	3 165 034	City
29	*We of the Never Never*	1982	3 112 000	Wilderness
30	*My Brilliant Career*	1979	3 052 000	Rural

Source: Australian Film Commission, 1998:222–4.
Note: *Total reported box office to 4 June 1998; some estimates were made by the AFC for films released prior to 1988.

that: 'It is not easy for workers in this industry to defy the pressure to contrive repeated exposures of conventional and safe Australianisms.' Canonic films were thought to be assured of domestic sales. These products were also reliable for returns from overseas sales, where audience comprehension of such narrow cultural constructions can be anticipated. Such films have been described as culturally bland, banal and internationally inoffensive (Crofts, 1989:131; Rowse and Moran, 1984:239). Films outside of this canon were thought doomed to being exhibited in a small number of theatres to small audiences.

Early exceptions to the 'Australian canon' were the films *The Adventures of Barry McKenzie* (1972) and *Alvin Purple* (1973). *McKenzie* depicted the exploits and experiences of Barry, a naïve, crude, disaster-prone Australian male (akubra wearing) in London. Similarly, Alvin was an urban dwelling, naïve innocent and, like Barry, he suffered from the uninvited sexual attention of women. The films portrayed the stereotype of the easy-going, beer-drinking, self-deprecating Australian 'good bloke' (Davidson, 1987:123–5). These constructions of the Australian male were also represented in the iconic film *Crocodile Dundee* (Davidson, 1987). Paul Hogan's character, Mick Dundee, was also an easy-going beer-drinking innocent abroad. However, Mick Dundee was a much more astute, entrepreneurial and less boorish character than was Barry McKenzie (Adams, 1986; O'Regan, 1988:164–7). *Crocodile Dundee* was an exceptional domestic and international success. It was not a canonic film and neither was it a simple parody of stereotypical Australian traits. *Crocodile Dundee* was a film that skilfully poached and utilised stereotypes of Australia, but without overtly criticising the stereotypes and avoided giving offence to the domestic audience (Abbey and Crawford, 1987:146; Morris, 1988). For instance, the viewer was shown how the lead character reproduces fictions and myths about bush life: including judging the time from the sun (but actually using a watch), shaving with a hunting knife (but actually using a razor) and preferring baked beans to goanna meat (Abbey and Crawford, 1987:146; Roddick, 1986:40). The film capitalised on stereotypes about Australian men, while also sending them up. Some wholly unreal and hackneyed Australianisms were simultaneously satirised and cherished (Roddick, 1986:40). In this way, the audience was provided with certain familiar traits, with which they could associate, but they were not overtly criticised. This skilfully allowed Mick Dundee to be one of us or an 'everyman' (Adams, 1986; O'Regan, 1988:164, 166–7).

Film reviews stated that Mick Dundee and the film in which the character appeared 'oozed Australian-ness' (cited in Abbey and Crawford, 1987:145). The film was always destined to 'provide a statement of Australia's cultural self-understanding' (Abbey and Crawford, 1987:145). A Sydney newspaper announced that:

> Australia has a new roving ambassador – the tough, laconic and phenomenally successful *Crocodile Dundee*, otherwise known as Paul Hogan . . . a plug for Australia – a positive, bright, breezy Australia bristling with energy and talent, not the whingeing, negative world-owes-me-a-living Australia some people seem to prefer (*The Mirror*, 29 October 1986).

Hogan himself stated that the character Mick Dundee was 'an attempt to give Australia a hero' (quoted in O'Regan, 1988:163). *Crocodile Dundee* was welcomed in early film reviews as Australia's first proper mainstream film (Roddick, 1986). Ten months before the film was released in Australia, its lead actor had promised an 'Australian fair dinkum film' that should 'make millions . . . and a pretty decent reputation for Australians as fair dinkum filmmakers' (Maddox, 1985:7). The film took a return of A$2 million in the first week of release and A$47 million throughout 1986. It became the biggest selling film in Australia and was only recently outsold by the American film *Titanic* (AFC, 1998:220; Roddick, 1986:40). The 600 investors in *Crocodile Dundee* are estimated to have reaped a 730 per cent return (Attwood and Helley, 1988:72). The extraordinary financial success added to the credibility of the film and, *ipso facto*, to the constructions of Australian-ness within it.

There has been a set of noticeable shifts in the landscapes represented within Australian feature films. *Crocodile Dundee* can be situated along this continuum of change. The canonic films mentioned earlier, particularly those of the 1970s (Table 3.1), were focused within rural locations. They were based on older literary pieces, rather than recent novels. Rowse and Moran (1984:242) concluded that 'productions have shrunk from the urban and favoured the rural'. Carter (1996) argued that Australian national identity has drawn heavily on landscape images, principally from the rural, and this is reflected in the Australian feature films of the 1970s. Settlement mythology, the bush myth, were composed of narratives of Europeans living, working, suffering and fitting into the land. The canonic Australian films included bush landscapes – but they were also peopled or being peopled – they were rural.

The dominance of rural landscapes in Australian film has been challenged by the centre or wilderness. These wilderness landscapes are ones in which (non-indigenous) people appear only as transients. *Crocodile Dundee* is an example of such a film. Earlier examples are the *Mad Max* films (1979, 1981, 1985) and more recently *The Adventures of Priscilla, Queen of the Desert* (1994). In the years since the 1988 Bicentenary of nationhood, Australians have increasingly identified with the 'red centre', the desert, 'Uluru', 'Kakadu' and other areas of wilderness. The prominence of the rural in feature films has waned:

> In the place of a landscape of bush or pioneer settlement that leads forward narratively into the present with the force of both evolution and destiny, there is instead an ancient and primeval landscape that leads backwards into the *pre*historic past (Carter, 1996:91).

These landscapes and the long standing human occupation of them have provided Australian national identity with a depth and centredness (Carter, 1996:91–2). Many reviewers noted how *Crocodile Dundee* included long scenes depicting 'outback' landscapes: 'much in the manner of a travelogue', 'outback travelogue', 'strikingly beautiful' (Abbey and Crawford, 1987:147; Crofts, 1992:216). For Carter (1996:90–1), the centrality of wilderness within *Crocodile Dundee* was an indicator that the bush myth was now too vulnerable to parody and was unable to be taken seriously. Both *Crocodile Dundee* and *Priscilla* are comedies, yet neither film parodies the wilderness landscape. Rather, it is the urban, especially the country town, that is the focus of critical representations.

In some senses *Crocodile Dundee*, like the canonic films, was anti-urban. Urban landscapes are portrayed as a site of problems and injustice, whereas the rural and wilderness are portrayed as harsh, but communal and level playing fields. Nonetheless, the film opens and finishes in the urban realm and Mick Dundee is able to reconstruct pockets of communion within the inner city. But this iconic Australian film lies at odds with the multi-cultural and highly urbanised nature of contemporary Australia (Abbey and Crawford, 1987:146; O'Regan, 1988:168). It is noteworthy here, that Hogan's third major film project, *Lightning Jack* (1994), was much less successful in Australia than *Crocodile Dundee* and its sequel. Perhaps the Australian cinemagoer had seen enough of both the bush and wilderness myths by the mid-1990s.

There is some evidence that urban landscapes are beginning to figure more overtly in Australian feature films and that both the rural and wilderness landscapes were reaching their 'use-by dates' (Carter, 1996:95). Australia has produced a number of moderately successful 'serious films' that are set within cities. They include *Shine* (1996), *Puberty Blues* (1981), *Malcolm* (1986), *The Sum of Us* (1994) and *Romper Stomper* (1992) (Table 3.1). These films explore problematic relations of gender, sexuality, disability and ethnicity within the city and provide social commentaries of one type or another. There have also been the more financially successful comedies that are set within the city, in which cultures of suburbia are satirised: *Strictly Ballroom* (1992), *Muriel's Wedding* (1994) and *The Castle* (1997). These films poke fun at suburban life, but they also portray comfortable and happy suburbanites (Mee and Dowling, 2000). Films set in suburbia that portray a problematic and meaningless existence, such as *Idiot Box* (1997), have much more limited success at the Australian box office. Films like *Idiot Box* may be critical successes, received by middle-class film critics as 'social realist' films or authentic representations of working-class life in the suburbs (Mee and Dowling, 2001). However, the film may not reflect the lives and interests of those people residing in the landscapes depicted. Mee and Dowling (2001) demonstrated how *Idiot Box* fitted the frames of reference of middle-class film critics from the inner city, but it did not fire the interest of the mass of filmgoers who reside in the suburbs.

Seeing the 'self' in film is clearly considered important, but determining how the 'self' should be seen is highly contested. The critiques of *Crocodile Dundee* provide an insight into the politics and ideology of such debate. Three types of *Crocodile Dundee* critic can be identified and each comes from a particular perspective, with their own set of ideas about Australian-ness and the role of film. First, *Crocodile Dundee*, and Hogan himself, were criticised by what could be broadly called the cultural elite, for portraying and celebrating Australia as a backward and 'outback' place populated by coarse, vulgar, philistines (see O'Regan, 1988:167). For example, Adams (1986) reported on indignant letters to newspapers that had railed against Hogan's being awarded the Order of Australia. For these critics, it is important to present Australia as an urbane and urban society. The second set of detractors was film critics who were disappointed at the superficial nature of the film and its lack of depth of interpretation. O'Regan (1988:157) observed that the film 'was hardly in the main swim of Australian feature film output'. In other words, the film was not within the canon described earlier. For Adams (1986), *Crocodile Dundee* was too slow, poorly edited, had a stingy musical score and inhibited camera work. In other words, the film was not an artistic product in the way that many of the canon had been. Nonetheless, the film was a stunning economic success.

A third set of critics of *Crocodile Dundee* focused on the conservative politics within the film. Cultural and social critics pointed out that the lead character was dismissive of most forms of radical politics and social reform and implied a very neo-liberalist politics in which rugged individuals should look after themselves (Brady in Doogue, 1987:3; Davidson, 1987:125, 127; Morris, 1988:114–5). Indeed, Hogan himself had utilised anti-government rhetoric and even professed the benefits of a benevolent dictatorship (O'Regan, 1988:161–2, 166). Mick Dundee was a non-urbanite transgressing the unfamiliar and unjust city, but unlike earlier rural innocents, he was not there to root out corruption or to advocate wholesale reform (O'Regan, 1988:166). In *Crocodile Dundee*, feminism was dismissed as an outdated irrelevance and the female lead character, a globe-trotting journalist, was reduced to dependency on the brave male lead (Abbey and Crawford, 1987:149). Reviewers noted with disdain how the female lead was objectified and ogled at (Crofts, 1992:221; Roddick, 1986:40). Issues to do with Native Title in Australia were sidestepped through some folksy logic regarding the futility of fleas arguing over the dog they infest (Carter, 1996:92; Davidson, 1987:125). The knowledge and expertise of indigenous Australians were unproblematically integrated and deployed by the white male, sidelining Aboriginals, and both clumsily and skilfully questioning their relevance and authenticity (O'Regan, 1988:170–1; Roddick, 1986:40). Morris (1988:115) commented that the film was overtly 'concerned with legitimating "majority" opinion'. For this third type of critic, film should trouble, and not legitimate, majority opinions that reinforce oppression and dispossession. This would require critical attention being drawn to problematic aspects of Australian culture.

If Abbey and Crawford (1987:152) were correct in their assertion that national cultural 'self-understanding will be crucially influenced by the way in which our artistic products portray our environment and social context', then Australian feature films have mostly constructed a bush or wilderness national mythology. Only in recent times has there been much evidence that film is representing the 'varied aspects of [Australia's] own traditions'. This may go some way to explaining why official determinations that Australia is a multicultural nation (Office of Multicultural Affairs, 1989) have made a very limited incursion into the popular consciousness. And in the context where overseas sales, specifically to the USA, have become central to the economic success of a film (see Box 3.2), cultural specificity and complexity are likely to be frequent casualties (Abbey and Crawford, 1987:150; Crofts, 1989:141).

Box 3.2 The international consumption of *Crocodile Dundee*

The film *Crocodile Dundee* was immensely successful in Australia, but it also enjoyed considerable overseas sales. The sales over the 12 months after release in the USA were US$174 million, compared to A$47 million in Australia. The film became the best selling foreign film in the USA ever (Crofts, 1989:134). The sequel film was financed and distributed through the Hollywood system and opened in 2837 cinemas across the USA, reaping US$29 million in the first six days of exhibition (O'Regan, 1988:157). The film was also a great success in the United Kingdom, France, Japan and Denmark, but it was released in these markets after 'busting blocks' in the USA and therefore carried a substantial marketing momentum (Crofts, 1989:141; 1992:223).

The success of *Crocodile Dundee* was all the more remarkable given the extremely poor performance of Australian films in the American market. For example, the Australian film industry produced an average of 28 feature films a year in the 1980s and 1990s, yet only an average of eight achieved a screen release in the USA (AFC, 1998:76, 121). One of the reasons for this is the tight control that Hollywood production companies have over the distribution and exhibition of film in the USA (Crofts, 1989:130). Another, and more powerful, explanation is the cultural tastes of the American cinemagoer. To put it simply, there is a disinterest in Australia, which reflects a general cultural insularity (Crofts, 1989:129, 132).

Crocodile Dundee was able to work around this insularity in three significant ways. First, the lead actor, Paul Hogan, had already established a substantial presence in the USA through his Australian Tourist Commission and Fosters beer advertisements and through his television comedy specials (O'Regan, 1988:161). The ATC adverts – 'Put another shrimp on the barbie' – had run in eight American cities. The trade journal *Encore* estimated that Hogan had benefited from as much as US$20 million worth of marketing exposure (Maddox, 1988:22). The Hollywood production company, Paramount, also fully assisted the distribution of *Crocodile Dundee*, opening in 500 cinemas across the country and touted it as being as good as a Hollywood film (Crofts, 1989:133–4; 1992:217). The film benefited from a level of exposure and marketing that other Australian films do not.

Second, the film was altered for its American release. This included substantial editing out to enhance the pace of the film to suit the taste of the American consumer better (Crofts, 1989:134–40). These cuts came principally from long segments on the Australian outback. A whole range of dialogue was changed or excised in order to avoid any unfamiliarity for the American viewer. Australian vernacular, such as stickybeak, billabong, g'day, strewth, stone the crows, bastard and fucking, were removed. These cuts removed 'Australian slang and less becoming behaviour in the interests of a more WASPish audience' (Crofts, 1989:140).

Third, the film story was set in New York, as well as at Walkabout Creek (Northern Territory) and the female lead was an American, thus providing
(continued)

(continued)

American viewers with a familiar association. Furthermore, the film studiously avoided any criticism of American culture and in this way it proscribed the provision of offence to mainstream audiences. Indeed, Abbey and Crawford (1987) have argued that the film said more about contemporary America than it did about Australia and all that was said was positive. Indeed, all the 'problems' in American society are resolved in the film – 'Blacks are in their place (as servants and attendants), women are in theirs, and even street terrorists are vanquished' (Abbey and Crawford, 1987:149–50). British reviewers noted how the film flattered and idealised both Australian and American society (Crofts, 1992:220). The wilderness landscapes in Australia were read as evocative of a 'smogless, egalitarian American heaven on earth', in which American frontier myths and individualism reign (Crofts, 1992:218–19).

The overseas success of *Crocodile Dundee* had a definite impact on inbound tourism to Australia. Tourism Australia officials noted the specific demand from Americans during 1987 to visit Kakadu country (cited in O'Regan, 1988:172–3). Resorts and events in north Queensland and the Northern Territory were constructed along a crocodile theme, some with a crocodile shape. Some promised the opportunity to swim with and eat these 'ferocious beasts' (O'Regan, 1988:173). However, the stock market crash and recessions tempered this rush to consume *Crocodile*.

3.3.2 Musicscapes

In as much as Australian film embodied debates about Australian-ness and negotiated between economic strategy/profit maximization motives and the desire for film products to contribute to the construction of national identity, music as a cultural text can as well play a central role in a nation's (or state's) ideological attempt at constructing an identity around which to rally people. The example of Singapore is a case in point.

Singapore is primarily constituted of a migrant society. As an independent state, it dates only from 1965 when it gained independence from British colonial rule. In two to three short decades, it has made the leap from a poor developing country to one enjoying First World living standards. Yet, while the GDP records strong performance and despite significant harmony among its multiracial population, it still has some way to go in terms of its constitution as a 'nation' of people with a strong shared history and destiny (see Kong and Yeoh, 2002). The government is active in encouraging this sense of identity and belonging and among its strategies in 'total defence' is an effort at building 'psychological' and 'social defence'.[1] This entails, *inter alia*, using a form of popular culture that people enjoy and identify with – music.

In 1988, the Psychological Defence Division of the Ministry of Communications and Information developed a 'Sing Singapore' programme in which 'national' songs were commissioned and/or encouraged. It produced various texts, including a *Sing Singapore* book containing the lyrics and scores of 49 songs, cassette tapes and a series of video clips aired on national television. While these songs do not represent popular music as the term is commonly understood, they are 'popular' in the way in which they are a part of the everyday lives of many Singaporeans. For example, these songs are taught in schools and are aired on national television and radio. Many young children learn them in school and they have become so much a part of the children's learned culture that Dick Lee, local composer and artiste, has commented that these national songs could well develop into the folksongs of future generations (Kong, 1995).

The national songs as put together in the 'Sing Singapore' package (including the lyrics and rhythms), the music videos aired on national television and the various timed releases of some of these songs are all combined with the intention of achieving an ideologically hegemonic effect – the construction of a particular version of a 'nation' and 'national identity'. Through various means of dissemination (including the constant airing on national television and radio; the organisation of community singing sessions in community centres; and teaching schoolchildren these songs during school assembly time at the directive of the Ministry of Education), the attempt is to persuade and reinforce in Singaporeans the idea that Singapore has come a long way since its founding (in 1819) and independence (in 1965); and that Singaporeans must play their part in continuing this dramatic development. In all these, the ultimate concern is to develop in Singaporeans a love for their country, a sense of patriotism and a willingness to support the ruling elite who have led the country through the short years since independence to tremendous development. As Dr Yeo Ning Hong, then Minister for Communications and Information, wrote in his message for the *Sing Singapore* songbook:

> Singing the songs will bring Singaporeans together, to share our feelings one with another. It will bring back shared memories of good times and hard times, of times which remind us of who we are, where we came from, what we did, and where we are going. It will bring together Singaporeans of different races and backgrounds, to share and to express the spirit of the community, the feeling of togetherness, the feeling of oneness. This, in essence, is what the 'Sing Singapore' programme is about (*Sing Singapore*, 1988:no p.).

Evidence of the state's hegemonic intentions in the construction of 'nation' can be found in the lyrics of the national songs. For example, Singaporeans are encouraged to express feelings of love and pride for and belonging to their country. For example, in 'This is My Land', Singaporeans are encouraged to proclaim that

> This is our island
> O Singapore, We love you so, love you so (*Sing Singapore*, 1988:116).

In turn, pride for the country is expressed in 'Sing a Song of Singapore':

I want the world to know about my island in the sun
Where happy children play and shout, and smile for ev'ryone (*Sing Singapore*, 1988:109).

Feelings of love, belonging and pride must, however, be translated into more active manifestations of citizenship and in the national songs, Singaporeans are exhorted to attain excellence for Singapore. This idea of excellence is bigger than, but encompasses various other concepts such as unity, commitment to Singapore, productivity, hard work and teamwork. The message in brief is this: if Singaporeans can ensure that they have these qualities and mindsets, excellence can be achieved for the country and indeed ensure that Singapore stays ahead of its competitors.

To encourage Singaporeans to play these various roles and to help attain excellence for Singapore, national songs appropriate past and present conditions. Specifically, past achievements are glorified; the role of the government is exalted; and both the built and natural environments are appropriated to extol Singapore's beauty and achievements. In highlighting these achievements, Singaporeans are encouraged to continue to give of their best for their country, to defend it and to give support to the ruling order.

The appropriation of the past is designed to remind Singaporeans of how successfully the state has been steered from struggling Third World conditions to newly industrialised status and to arouse a sense of pride and loyalty. For example, in 'We are Singapore' (*Sing Singapore*, 1988:95), Singaporeans are reminded that:

There was a time when people said
that Singapore won't make it
but we did
There was a time when troubles
seemed too much for us to take
but we did.
We built a nation strong and free.

Through this reminder, Singaporeans are encouraged to think that no problem would ever be too difficult to handle as long as Singaporeans continue to uphold the spirit of the pioneers.

Part of Singapore's successes, as the songs remind us, could have been achieved only because of good government. The Tamil songs 'Engkal Singapore' (Our Singapore) and 'Paduvom Varungal' (Come, Let's Sing), underline the view that Singapore has a 'strong and effective' government that 'cares about the welfare of her people' (*Sing Singapore*, 1988:115, 97). The result of such good government is that 'life is joy and harmony' ('Voices from the Heart', *Sing Singapore*, 1988:100). Music is thus used as a tool to help legitimise the ruling elite. Indeed, to reflect the good life that Singaporeans are said to enjoy, elements of the built and natural environment are appropriated to conjure images of progress, peace, joy and harmony. For example, in 'Sing a Song of Singapore', happy children are said to be playing:

In city streets in parks so green in highrise housings [sic] too
In every place so fresh and clean On sunny beaches too (*Sing Singapore*, 1988:109).

In those two lines, the urban environment of 'buildings . . . climbing all the way to the sky' ('Singapore Town', *Sing Singapore*, 1988:100) and the rustic peace of (limited) natural environments in Singapore are appropriated for ideological ends, namely to remind Singaporeans that if they did not continue to support the status quo and strive together with their leaders towards excellence, they stood to lose the beauty and prosperity of their 'fair shore' ('The Fair Shore of Singapore', *Sing Singapore*, 1988:110). The images of skyscrapers are particularly important as symbols of modernisation and development as well as of triumph over an environment that offered nothing towards economic progress by way of natural resources. At the same time, the emphasis on the 'fresh', 'clean', 'green' and 'sunny' can work at two levels. On the one hand, it represents a recognition of the value of the natural alongside the developed, reflected in the fact that the lyricist was determined to recover 'sunny beaches' as part of Singapore's natural heritage when in fact, few natural beaches remain, given the extent of land reclamation and seafront construction over the last two decades. If read at this level, the tension between urban development and conservation of the natural environment that has been vocalised in public arenas (see, for example, Savage and Kong, 1993:42–44) has been glossed over, whether in innocence or for ideological ends. On the other hand, the portrayals may also represent a proud acknowledgement of the consciously created manicured parks in 'clean and green' Singapore, this time, glossing over the ironical fact underlying the creation of the 'Garden City': the many manicured parks and 'instant trees' could only be created and planted by rolling back the natural heritage that had existed before (see Kong and Yeoh, 1996).

While popular music is drawn on to construct versions of the 'nation', at times, appropriating through the songs various natural and built landscapes in Singapore to underscore the messages that the state wishes to emphasise, the same cultural form can be reappropriated by others to express counter-constructions. Kong (1995) used the framework of parody in popular songs to analyse resistance to the ugly side of the Singapore 'national identity'. Referring to a volume entitled *Not the Singapore Song Book* (1993), which contains new lyrics set to popular tunes, including some national songs, she illustrated how many of the lyrics are couched in a tongue-in-cheek tone and are written with a dose of humour, often parodying national songs and Singapore life. Yet, she argues that the humour belies more deep-seated concerns of the lyricists, which address government policy on the one hand and the 'ugly Singaporean' on the other.

As one example, she illustrates the way in which the state's concern that Singapore's fertility level is below replacement rate has resulted in policies which exhort Singaporeans to reproduce. In particular, the government message is directed at women and highly educated women especially, because of the belief that they produce more intelligent children, thus maintaining the quality of the gene pool. This has been a subject of controversy in Singapore and is reflected in Ong Cheng Tat's tongue-in-cheek 'Count! Mummies of Singapore' set to the tune of the national song, 'Count on Me, Singapore'. While the original lyrics exhorted Singaporeans to stand up and be counted among those who would give their 'best and more' to the country (*Sing Singapore*, 1988:95), Ong's version translates the exhortation for Singapore women, urging them to reproduce. Ong's reminder to women is that:

> We have the ova in our bodies,
> We can conceive,
> We can conceive.
>
> We have a role for Singapore,
> We must receive,
> We must receive (*NTSSB*, 1993:35).
>
> There's a spirit in the air,
> Telling us to be a pair!
> We're going to get our hubbies, start a family . . .
> We must conceive!
> We MUST conceive!! (*NTSSB*, 1993:35).

In the same vein, Ee Kay Gie writes in tongue-in-cheek fashion lyrics that are to be sung to 'The Colonel Bogey March', exhorting Singaporean women to get married for the sake of the nation:

> Hey girl!
> Why aren't you married yet!
> You girl!
> A man's not hard to get!
> . . . .
> Now's the time for you to choose your mate!
> Don't delay! Do not procrastinate!
> Wed now and do your nation proud –
> Then do your part, spread the word clear and loud (*NTSSB*, 1993:5)!

Apart from the lyrics contained in the book and views expressed by the lyricists, a live skit was staged, a combination of chat, acting and song served to bring to life the lyrics inscribed in *Not the Singapore Song Book*. Held at a local discotheque that is known to be the watering hole of undergraduates and other young working adults (often yuppies), the skit ran for seven days in 1993. The choice of this particular landscape of popular culture in which to share the satire and parody is redolent with symbolic meaning. First, it represented a landscape of fun and entertainment, antithetical to the values of hard work and gainful employment the state wanted to inculcate. Second, it was a place of youths, the group that the government wanted badly to reach out to and 'convert'. That the skit played to audiences mainly in their twenties, comprising that category often identified by the government as the generation that is more highly educated, more vocal and more critical of government policies, was significant. In as much as landscapes in popular culture convey socially constructed meanings, landscapes of popular culture, or popular culture in landscapes, are also invested with symbolic meanings.

3.3.3 Foodscapes

While the case of Singapore just discussed illustrates the presence of the state as either directly and plainly stark or palpable as a presence ordinary people respond to, the relationships between everyday landscapes and national identities need not always be mediated by the state. Take the example of food or cuisine as part of the cultural landscape. Cuisine is a key indicator of national identity. In some long established nations, cuisine forms part of their histories and identities. For example, British cuisine is often identified as 'meat and two veg', with numerous variations on this including roast beef, soggy cabbage and heavy steamed puddings. Cuisine may be used as a marketing ploy, to promote gastronomic tourism, as in the case of Scotland which promoted a much richer food heritage including luxury items such as salmon and venison, making use of discourses of authenticity and heritage (Bell and Valentine, 1997).

Many newer nations have had to work hard to define a clearly marketable culinary heritage and gastronomic landscape. This has been the case in Australia, which inherited an Aboriginal use of foodstuffs that was largely ignored and derided by the European occupants for two centuries. The disregard of the Aboriginal sources of food meant that the infant colony was often on the verge of starvation and individuals, such as the explorers Burke and Wills, perished while indigenous populations flourished around them. It is only in the last ten years that the Australian 'bush tucker' has become an item to be exoticised, with emu, kangaroo and crocodile finding favour in up-market restaurants (Ripe, 1996).

With British settlement came British food: meat and two veg, Lancashire hotpot, heavy steamed puddings and dishes that later were claimed as authentically Australian, such as meat pie with tomato sauce. Many of the dishes and traditions imported to Australia were particularly inappropriate to the environment, such as full-scale Christmas dinners and puddings baked and steamed in searing summer heat. Australian cuisine has subsequently been developed and extended by more recent waves of immigrants, bringing with them the cuisine of the Mediterranean, the Middle East and Asia. As a consequence, Australian culinary habits have changed dramatically and the restaurants have reflected the growing fusion of culinary styles.

In 1994 two Australian chefs were refused work visas to enter Japan on the grounds that there was no such thing as 'Australian cuisine' (Ripe, 1994). This decision was overturned a few days later but it caused a considerable amount of discussion in the Australian media. Even if there is not an 'authentic' Australian cuisine based in a peasant economy and local foods, Ripe (1994:15) argues that there is:

a distinctive, Australian style in our food which has nothing to do with lamingtons or Vegemite, meat pies or sausage rolls, pavlova or peach Melba. We have potentially the most exciting and eclectic cuisine in the world developing here, one that borrows from everywhere. Australian cuisine is defined by its very eclecticism.

It is not defined by its use of kangaroo and emu or even macadamia nuts (one of Australia's very few indigenous foodstuffs to be turned into a commercial crop), but essentially by its incorporation of Asian ingredients into western cuisine. In a flight magazine in 2001, Geoff Lindsay, chef at a Melbourne restaurant was quoted as saying:

> I am a fifth generation Aussie boy who is seduced by ginger, chilli, palm sugar, Turkish delight, chocolate and pomegranate. My food is intrinsically Australian, and I am proud of that. It is multicultural and layered with influences from all over the world, and spiced with new frontier enthusiasm (Pickett, 2001:106).

Lindsay uses his words carefully. He is an 'Aussie boy', with connotations of the dinky-di Aussie from the bush, who has been 'seduced' by the exotic food of distant lands (notably the east). Yet he defines this mixture as simultaneously 'multicultural' and 'intrinsically Australian', implying a changing notion of national identity, which is still enlivened by the new, using the metaphor of the frontier. The development of this cuisine has had a significant impact on the gastronomic landscape, with the changing landscape reflected at home, in individual eating habits, in supermarkets and, most visibly, in restaurants.

Restaurants in Australia have become particularly multicultural with a huge range of European and Asian influences. A survey of the restaurants in a long standing Italian migrant suburb of Newcastle, New South Wales, found that while half the restaurants were Italian, there were also Yugoslavian, French, Greek, Dutch, Lebanese, Thai, Vietnamese and Chinese, as well as multicultural Australian. The restaurants in Newcastle's Beaumont Street are marketed as 'Eat Street'. They form a distinctive gastronomic landscape which reflects the nation's waves of immigration and attract tourists and locals alike.

Through the examples of filmscapes, musicscapes and foodscapes, we have illustrated in this section the roles of cultural landscapes in the constructions and contestations of national identities. This has been achieved through the selective use of landscapes in texts, the constructions of particular landscapes within texts (including hybrid ones) and the presentation of alternative landscapes as contestations within texts.

3.4 Everyday landscapes of local identities and communities

Our earlier discussion of cultural globalisation, imperialism and homogenisation underscore the interconnectedness of the global and local in the construction of everyday landscapes. Here, we turn the discussion towards a focus on the local and draw attention to the fact that, the power of the global notwithstanding, there are those who argue that cultural forms are necessarily local in their production, consumption and effects. In an earlier section, we illustrated how transcultural forms of music and food illustrate that, in the production of cultural forms, local influences often infuse global forms, thus asserting local identities.

3.4.1 Musicscapes

We turn attention here to the assertion of the local in the consumption of music and the concomitant effects. Smith (1994:237) argues in the context of music that music can be a form of resistance to the homogenising forces of the culture industry, 'not necessarily by producing an alternative sound, but by enabling people to experience music in distinctive localised ways'. Relatedly, Street (1993:54) argues that participation in local musical activities can contribute to people's sense of identity. Drawing on the works of Cohen (1991) and Finnegan (1989), Street (1993:54) holds that:

> [L]ive music, because it is necessarily local, being available only in a specific place to a limited audience, is particularly effective at serving a sense of community identity. It serves to differentiate those consumers from others, while simultaneously locking them into national trends and events.

In a slightly different vein, others have argued that participation, whether as performers, audience or supporters in local musical activities, promotes community identity. In nineteenth- and early twentieth-century Britain, for example, brass bands were a means of forging a sense of community and locality (Herbert, 1992). As Russell (1991) points out, they were closely linked to the ceremony and ritual of public life of a particular locality and thus became bound to the rhythm and identity of that community. In addition, as Bevan (1991) illustrates, brass band competitions whipped up a sense of loyalty for the local 'team', thus fostering community identity.

In focusing on a seldom studied phenomenon, Kong (1996b) drew attention to *xinyao* in Singapore and its role in developing a sense of youth and local community and identity. *Xinyao* is a shorthand term for *xin jia po nian qin ren chuang zuo de ge yao* ('the songs composed by Singapore youths'), an extraction of the first and last words of the longer term. It began around 1980 with groups of youths in junior colleges[2] who wrote their own songs and played their compositions informally for one another within the confines of school and during their leisure time. Apart from students, some were also fresh graduates from junior colleges, serving national service,[3] entering university or in their first jobs. As one observer had it, they were mainly 'boys and girls next door' singing about their feelings. Many did not have any prior musical training. Their compositions thus generally consisted of simple melody lines using basic chords and, particularly in the early stages of *xinyao* development, it would not be uncommon to find that 'two guitars at most provided the rhythm' (*The Straits Times*, 22 December 1985). Occasionally, there would be some supplementary piano, flute or violin accompaniment.

The early *xinyao* groups were isolated groups of teenagers with a love for composing and singing, sharing their music informally within their schools in their own time. Few had any sense that there were others with like interests.

Gradually, as these groups began to organise and host joint activities, their participation as composers, performers and organisational personnel gave rise to a sense of community identity. Kong (1996b) cites a performer in the early years of *xinyao* expressing her sense of connectedness with others, the sense of common interest, a shared strength and purpose. The performer, Pan Ying, had this to say:

> I feel that it was quite good then. Maybe that period coincided with the fact that we had just left school and so we were still very nostalgic about school life. Maybe because the group of friends were very innocent. Our relationship was very good. I feel that there was a kind of strength. Everybody worked together. Although the performance was only once a year, everybody treated it with the utmost import-ance. Some did the props, some did the control room duties, some were respons-ible for food etc. Everybody performed their duties well.

As *xinyao* further developed, groups were formed at community centres.[4] *Xinyao* courses were conducted at some community centres, which provided support by allowing free use of their facilities and musical instruments for practices, as well as help in organising concerts. *Xinyao* camps were also organ-ised. They involved the exchange of ideas regarding a range of issues, includ-ing the improvement of skills, musical trends and the role of *xinyao* in social functions. Each of these centres developed communities of *xinyao* enthusiasts in their own rights, with each meeting regularly, practising and sharing their music. Very real local communities were therefore evolving. These groups would also be invited to perform at local community events, for example, at community centre functions such as National Day dinners and at neighbour-hood shopping centres, thus becoming bound to the rhythm and identity of that community (Kong, 1996b).

While musical activities can contribute to the spirit and identity of local community and landscapes, other popular local practices are similarly markers of identity, sometimes through the process of exclusion and delineation. We use the examples of consumption of food and the practice of graffiti to make the case.

3.4.2 Foodscapes

The consumption of Filipino food in Hong Kong is, as Law (2001:280) argues, 'a salient example of how everyday experience can become a performat-ive politics of ethnic identity'. Along with music, photographs and letters, food contributes to the transformation of Central Hong Kong into Little Manila each week. Law (2001) recounts how, every Sunday, Filipino domestic workers gather in and around Central Hong Kong, particularly in Statue Square, to eat Filipino food, read Filipino newspapers and magazines and consume a range of Filipino products from specialty shops. A buzzing com-munity develops on the streets and pavements where the ritual gathering occurs, giving the place a weekend identity, although not one welcome by all in Hong Kong's multicultural society. To quote Law (2001:276): 'The

Philippines is experienced each Sunday through a conscious invention of home – an imagining of place through food and other sensory practices that embody Filipino women as national subjects.' The taste and aroma of familiar Filipino food helps to create a landscape of familiarity, that of the Filipino home prior to migration. Critically, it is not any food, but home-cooked Filipino food, for it is that which creates a sense of place and causes a transformation of Central into Little Manila. It is also that which allows nation and ethnicity to be resignified, for these foods become 'national dishes' by their appearance in Hong Kong. Their consumption, Law (2001:278) argues, is infused with politics.

While a vibrant community and sense of identity is evoked in the particular locality of Central, tensions appear as others view Central to have become a place of 'filth' and 'stench' on Sundays. This disagreeability with the olfactory geographies of the locality is mostly about a rejection of the public presence and congregation of a group of women – Filipina domestic workers.

Such tensions also appear in the Hong Kong home between Chinese employers and Filipina domestic workers. These too are expressed in terms of contestation over food. Some domestic workers are not allowed to cook Filipino food at home because of 'unsavoury' food odours. Yet, as Law (2001:278) points out, 'claims about food odours are impossible to disentangle from ethnic stereotypes'. Power within households determines whether such food may be cooked in the Chinese kitchen.

In sum, the consumption of Filipino food by Filipina domestic workers in a particular locality in Hong Kong is instrumental to the creation of a 'home culture', even if it faces resistances from those who view such public congregation with disdain. At the same time, the cooking and consumption of Filipino food in the Chinese home by Filipina domestic workers is not always approved of and reflects the refusal to allow expression of Filipino identity at the very local level of the domestic. This reflects the construction of certain cultures as 'out of place', with Filipino cultural landscapes deemed inconsistent with dominant national ideology.

3.4.3 Graffiti

Graffiti are also an aspect of the everyday landscape that has diverse roles and causes. In some places, graffiti are almost omnipresent, yet in other societies they are rare and therefore particularly subversive. Graffiti have sometimes been used for outpourings of grief and to advocate peace (see Klingman *et al.*, 2000). They are simultaneously an indicator of vandalism, crime and malaise, as well as a form of expression, art and rejuvenation (Bowen, 1999; Gross *et al.*, 1997). Scholars from a range of disciplines have focused attention on graffiti, including researchers from art criticism, urban planning, sociology and law (see Bowen, 1999:22). The key geographical works have included Ley and Cybriwsky's (1974) analysis of the rise of graffiti in North America in the 1970s and Cresswell's (1992) examination of the incorporation of graffiti into the establishment world of art galleries from the 1980s. These geographical analyses have drawn attention to the politics and social processes associated with graffiti.

'Graffiti' is an Italian word that can loosely be translated to English as 'scratchings' or 'writings'. Graffiti practitioners refer to themselves as 'writers' (Kan, 2001:19). Graffiti take many forms. Some of the more famous examples of graffiti have included political forms, such as unification messages on the former Berlin Wall and pro-Sandinista slogans in Nicaragua (Ferrell, 1995:77). Some of the most well-known murals are those in Belfast, Northern Ireland (see Chapter 5). Animal rights and feminist activists use graffiti directly to engage the messages within billboards and other forms of outdoor advertising (see Waitt et al., 2000:463–5). There are also the private types of graffiti, such as 'doodling' and 'latrinalia' (see Table 3.2).

Table 3.2   A typology of graffiti

Type	Form	Purpose
Doodling	Scrawls and scribbles, as seen on school desks, etc.	Expression of scattered thoughts, divided and distracted attention, adolescents mostly
Latrinalia	Names, insults and jokes within public toilets	Expression and discussion of repressed ideas, emotions and politics
Tags	Sub-cultural nicknames or symbols: written in a bold and simple form in prominent locations	These tags place the mark of the writer on visible places where they can be read by their peers. The more prolific and visible, the greater the recognition and fame generated
Gang graffiti	Gang markers placed at the limits of gang territory	These can take the form of tags as gang identifiers rather than individual markers. They warn other gangs about the limits of territory
Political murals	Pictures and text portraying political events and messages	These murals protest oppression or advocate specific political initiatives (see Belfast murals discussed in Chapter 5)
Activist graffiti	Text painted on outdoor advertising images. These include billboards, hoardings and bus shelters	The graffiti take issue with an ideology or message implicit to the original image. This direct action problematises the image, drawing attention to its being contested
Pieces	Complex pictures utilising both text and graphics and covering a substantial space	Piece is a colloquial abbreviation for 'masterpiece'. They can be legal or illegal. Style and story are the quality criteria of assessment by peer muralists
Post-graffiti	Pieces painted within galleries or on canvases	Commissioned works for art galleries or buyers. Sold for subsistence, or to support the underworld lifestyles of writers

Source: Abel, 2000; Bowen, 1999:22–3; Ferrell, 1995:77–8; Gross et al., 1997:275–80; Kan, 2001; Lachmann, 1988:229–34, 239–40, 245–8; Powers, 1996:137.

Ley and Cybriwsky's (1974) analysis of graffiti in Philadelphia drew attention to the territorial role of graffiti. Wall marking was a means of defining turf or areas of control. Teenage street gangs, which are otherwise powerless, were able to use graffiti to assert claims on urban space and mastery (Ley and Cybriwsky, 1974:494–6). Lachmann (1988:239) found that street gangs would often employ local bombers (graffiti or tag writers) to perform this definitional role:

> Such graffiti are commonly boundary markers; they delineate an interface, the edge of socially claimed space, a boundary which in the North American city is often ethnic (Ley and Cybriwsky, 1974:501).

The text in this form of graffiti warns away intruders, often using code, such as simple alphabets written backwards to speak to both friend and foe (Kan, 2001:19). Ley and Cybriwsky (1974) were convinced that graffiti in the urban landscape could be read as an indicator of areas of ethnic tension. The extent of gang graffiti writing, and rewriting, could be used to identify areas of inter-communal hostility, and perhaps diagnose or predict future sites of conflict (Ley and Cybriwsky, 1974:501–5).

Political graffiti have inspired the punitive attention of state authorities. These have included the graffiti death squads of Nicaragua (Ferrell, 1995:77), and also the gaoling of Republican muralists in Belfast (see Chapter 5). In the everyday space of the city, transit police and other urban authorities have long running campaigns with bombers and crews. The policing costs in some North American cities have been huge, as have the cleaning costs for local councils (Ferrell, 1995:78). As much as US$4 billion may have been spent by US cities on graffiti cleaning in 1994 (see Kan, 2001). New technologies and procedures for surface coating, surveillance and cleaning agents have been brought to bear and some analysts report reductions in graffiti as a result (Ferrell, 1995:80; Powers, 1996:40; Ward, 2001). The sale of spray cans to youths has been prohibited in many western cities (Ferrell, 1995:80; Ward, 2001). Graffiti have inspired a rage among urban elites. The editor of *American City and Country* commented on the uncivilised nature of graffiti writers:

> Those who indulge in graffiti (they call themselves artists, I call them self-indulgent twerps) claim a lineage that goes back to our cave-painting ancestors (Ward, 2001:4).

Elected officials in the USA have called for corporal punishment of convicted graffiti vandals, including various calls at times for caning, paddling, lopping off hands, shooting and castrating (Ferrell, 1995:80–1; see Box 3.3). These calls for violent responses are reflected in the reports by graffiti writers of the beatings they receive from police (Ferrell, 1995:80; Lachmann, 1988:238, 244).

The risk of capture, police beatings and the dangers of tagging prominent landscapes or moving objects are part of the attraction of graffiti writing. Graffiti writers report to researchers the adrenalin rush and excitement generated through tagging (Abel, 2000; Ferrell, 1995; Kan, 2001; Powers, 1996). Writers construct themselves as in a contest with police or with society

---

Box 3.3　Caning a US bomber in Singapore

In 1994, Michael Fay, an American teenager living in Singapore, was sentenced to six strokes of the cane and four months' imprisonment and also fined a total of S$3500 (US$2233) for spray painting and damaging 18 cars in Singapore, including switching their number plates, and keeping stolen street signs. The resulting nationwide debate in the USA focused on the sentence and drew in then President Bill Clinton, former President George Bush and a legion of editorial writers, commentators and talk-show callers.

The debate initially centred on whether the punishment was too harsh. It soon turned inward, focusing on the failure of the USA criminal justice system. Certain moral geographies developed through the debates and invited assertions about the superiority of particular value systems. Singapore was viewed as leading attempts by east Asians to define and assert their values against those of the west and some drew up moral geographies that segregated the world along cultural and ethnic lines. One commentator, writing in the *Washington Post*, for example, wrote that: 'The drama surrounding Fay . . . points up the continuing sense of cultural superiority many Asians (Chinese, in particular) feel to many foreign nations (America, in particular)' (*Far Eastern Economic Review*, 28 April 1994). In these debates, landscapes of urban order and strict judiciaries became associated with civilisation and superiority. Landscapes of disorder, such as graffiti, were markers of less civilised society.

---

generally (Lachmann, 1988:235). Many researchers have analysed how graffiti may be a resistance to ordered and privileged society because writers violate spatial controls (Ferrell, 1995:79, 86). Writers talk of the attempts to silence or snub them and of their resistance to that (Ferrell, 1995:81). Urban authorities do treat graffiti as an indicator of disorder that must be contained and removed (Cresswell, 1992:329, 335). Studies of neighbourhood disorder have found that people perceive graffiti as grounds to be fearful, more so than abandoned buildings (Ross and Jang, 2000:408; Ross *et al.*, 2001:584, 586–7).

Contrariwise, graffiti writing has also been found to have very constructive and positive functions. These include evocations of anti-segregation and peace as well as urban beautification. Ferrell (1995:85) pointed to the multiethnic composition of crews in Denver, USA, and how they formed alternative communities that transgressed segregation. The mobility of writers across segregation boundaries, both socio-economic and cultural, was a resistance to the spatial control that constrains urban lives:

> Graffiti writing not only confronts and resists an urban environment of fractured communities and segregated spaces; it actively constructs alternatives to these arrangements as well (Ferrell, 1995:83).

Figure 3.1   Graffiti as a form of urban beautification

The style of tagging and pieces is also cross-national or global. Writers perform a culture that is beyond their local and national circumstances (Gross *et al.*, 1997). The pieces and tags incorporate international fonts and themes as well as local references. Writers make connections with the resistances of youth in distant locations and they short-circuit borders and hierarchies (Gross *et al.*, 1997).

Graffiti have also been officially deployed as a beautification and urban renewal initiative (see Figure 3.1). Most research with graffiti writers has found that they respond positively to the option of doing officially sanctioned pieces (Ferrell, 1995:79). Bowen (1999:22) found writers in Toronto to be 'community minded and concerned about the inherent aesthetics of the city'. In some jurisdictions, convicted graffiti vandals have been compelled to do official pieces by way of penance (Abel, 2000).

Graffiti murals or pieces are clearly perceived by writers to be forms of artistic expression. Indeed, the term piece is a colloquial abbreviation of masterpiece. Various researchers have found from their interviews with writers that mural painting is a medium of expression (Abel, 2000; Bowen, 1999; Kan, 2001). Furthermore, mural painting provides a rare opportunity for youth to participate in the direction and appearance of urban space. A graffiti writer said to Abel:

> I'd love to meet Michelangelo, man. He did pieces. He did productions. Did you ever see the Sistine Chapel? Not just the ceiling, man – the walls. He must have done twenty figures there. The man was awesome. If Michelangelo was alive, I'd give him a bunch of cans. He'd do the best train ever (quoted in Abel, 2000:59).

During the 1970s and 1980s in North America, there were a number of hesitant attempts to embrace graffiti formally within the art establishment.

Graffiti were constructed as a form of primitive art, 'poverty art', that had developed organically within areas of deprivation and despondency (Cresswell, 1992:341; Lachmann, 1988:246). The art world's interest in graffiti was fickle and fleeting; works failed to maintain significant prices and exhibitions were few (Bowen, 1999; Lachmann, 1988:245–8; Powers, 1996). To the established art world, pieces lacked a sophisticated aesthetic. Graffiti canvases, removed from their social and legal origins, also lost the only respect and credibility they had (Cresswell, 1992:341; Kan, 2001; Powers, 1996). Cresswell (1992:339, 343) argued that the post-graffiti movement was an attempt to co-opt graffiti, a move by the urban elite to reorder the city.

Everyday graffiti demonstrate two important aspects of the cultural landscape. First, the cultural landscape is an expression of culture, of individuals, of cultural groups and also of political views. The same could be said of the Filipino spaces of Central Hong Kong. These Filipino landscapes and the *xinyao* groups provide a space for the maintenance of culture and community. Second, landscapes reflect power relations and debate. Graffiti are a threat to urban order and Little Manila is seen as inconsistent with Chinese Hong Kong. The politicised nature of the cultural landscape is explored in more detail in the following two chapters.

3.5 Summary

Cultural geographers in the Berkeley tradition had tended to focus their attention on the exotic and the antiquarian, rather than the everyday. Retheorised perspectives in the late 1980s and 1990s gave more focus to the popular, although also trained the gaze to the spectacular (e.g. world's fairs and mega-malls – see Chapter 4). In this chapter, we have centralised everyday people and everyday lives in everyday landscapes and shown how, in real and represented terms, landscapes of everyday popular cultures are infused with power relations, ideological constructions, resistances and contestations. These relationships are sometimes played out at the level of the local, but often there are intersections between the global, national and local. We have shown how landscapes of everyday popular cultures are globalised and homogeneous in some ways (the 'McDonaldisation' of the world) but that this is not unquestioningly and peaceably accepted everywhere. Oftentimes, local resources and adaptations become evident so that a process of transculturation is at work. Everyday popular cultures and landscapes are also appropriated or evolve into markers of national identities because they become pervasive practices and infuse the national consciousness and, sometimes, taken-for-grantedness. At yet other times, such practices and landscapes operate at the level of the local and provide opportunities for the construction of local identities and communities, which, in turn, may be contested and debated. While our focus in this chapter has been to centralise the everyday and the popular, we will turn in the next two chapters to concentrate specifically on the issues of power and resistance, when our gaze will then embrace both the ordinary and the spectacular.

3.6 Notes

1. Total defence comprises military, civil, economic, social and psychological defence and is designed to prepare the country for any eventuality. Military defence, perhaps the most obvious form of defence, involves constant upgrading of the military, both full time and reservist, in terms of improved weaponry, professionalism, intra-organisational attitudes and effective liaison with the public (Seah, 1989:957). Yet, military defence alone will not suffice in times of trouble. Civil defence forces are deemed necessary to maintain the internal administration of the country under adverse circumstances, by organising rationing and distribution of vital resources, evacuating people and collecting blood, for example. Likewise, adequate economic defence ensures that government, business and industry will be able to organise themselves such that the economy will not break down during or under threat of war. While military, civil and economic defence take care of the material realm of existence, also significant and most pertinent to this discussion are social and psychological defence. The former refers to the promotion of cohesion among Singapore's diverse groups so that external subversion through the exploitation of inter-communal tensions would be minimised while ideals are fully shared by all Singaporeans. Psychological defence is defined as 'the means of winning the hearts and minds of the people and preparing them to confront any national crisis' (Seah, 1989:956).

2. In the Singapore education system, students may move from secondary schools to a two-year junior college education, at the end of which they sit for the General Certificate for Education 'Advanced' level examinations. Junior college education is meant to be an intermediate training ground for students who wish to pursue university education subsequently. Junior college students are generally 17 to 18 years of age.

3. National service, that is, two-and-a-half years in the army, is compulsory for all Singaporean males when they turn 19, with some state-approved exceptions.

4. Community centres can be found all over Singapore. Each caters to a particular housing estate, offering recreation facilities and organising varied classes and activities, such as cookery, flower arrangement, chess and guitar playing.

Chapter 4

Landscapes of power

4.1 Landscapes of power and power of landscapes _____

Power and the domination it entails is multivalent, ranging from open command and authority, to veiled control via persuasive strategies, that is, the exercise of hegemony. Often, the latter, when successfully used is more effective, as those dominated adopt the ideological positions of the dominant and powerful and are subject to control without recognising it. Power may be exercised by a range of groups, from states to capital to social groups such as gender, racial and religious groups. The role of landscapes is frequently integral to the exercise of power. The direct control of landscapes for particular uses or non-use is apparent as a form of dominance as in the example of state closure of Tiananmen Square during several anniversaries of the 4 June demonstrations in 1989. The hegemonic role of landscapes, by way of contrast, relies on their naturalisation of ideological systems, made possible because of their dominance in everyday lives and their very tangible and visible materiality, making that which is socially constructed appear to be the natural order of things.

The concepts of 'ideology' and 'hegemony' are central to understanding the power of landscapes and landscapes of power. J.B. Thompson (1981:147) outlined three approaches to the understanding of 'ideology'. His preferred reference is to ideology as 'a system of signification which facilitates the pursuit of particular interests' and sustains specific 'relations of domination' within society. While acknowledging two further definitions, Thompson (1981) criticised the first ('the lattice of ideas which permeate the social order, constituting the collective consciousness of each epoch') for being over-generalised and the second (a 'false' consciousness which 'fails to grasp the real conditions of human existence') for being too narrow and pejorative. In *Landscapes*, we utilise Thompson's preferred definition of a belief or meaning intended to benefit certain interests and to determine power relations.

Gramsci (1973) argued that 'hegemony' is the means by which domination and rule is achieved. Hegemony does not involve controls which are clearly recognisable as constraints in the traditional coercive sense. Instead, hegemonic controls involve a set of ideas and values which the majority are persuaded to adopt as their own. So as to persuade the majority, these ideas and values are portrayed as 'natural' and 'common sense'. This is 'ideological hegemony'. Applied in the context of ruling political elites vis-à-vis the masses, for example, the masses' acceptance of the ruling group's ideology, hegemonically purveyed, gives the ruling group the power to shape the political and social system. Put another way, to stay in power, a ruling group must persuade people it is working for their general good. Members of the ruling group must persuade people to accept their definitions of what constitutes this general good. People must also be convinced that the ruling group's methods of attaining this 'public good' are the most natural, commonsensical ones. If policies and actions are supported, the power of the ruling group is uncontested. The most successful ruling group is the one which attains power through ideological hegemony rather than coercion. When hegemonic control is successful, the social order endorsed by the political elite is, at the same time, the social order that the masses desire.

One of the key ways in which power can be expressed, maintained and indeed, enhanced, is through the control and manipulation of landscapes and the practices of everyday life. In both urban and rural landscapes, the powerful social groups will seek to impose their own versions of reality and practice, effecting their ideologies in the production and use of landscapes, as well as dominant definitions of their meanings. What they produce are therefore landscapes of power, that is, landscapes that reflect and reveal the power of those who construct, define and maintain them. These could be landscapes crafted by the powerful state or by capital, often by males or heterosexuals, and by particular races (see also Chapter 6). Once constructed, these landscapes have the capacity to legitimise the powerful, by affirming the ideologies that created them in the first place. This is achieved through their naturalising role. As Duncan and Duncan (1988) argue, landscapes naturalise ideologies and social realities because they are 'so tangible, so natural, so familiar . . . unquestioned'. In other words, they not only reflect and articulate ideologies and social relations, they actively institutionalise and legitimise them by reifying them in concrete form, thus contributing to the social constructedness of reality. They therefore contribute to the social construction of ideologies – often, of racial and gendered ideologies – and analysis of landscapes from this perspective foregrounds the social constructedness of categories through landscapes. Understanding this then focuses attention on the nexus between the cultural and political and spotlights the argument that power relations do not simply involve political and economic coercion/resistance but also ideological and cultural impositions/oppositions which are often inseparable from the material. In Chapter 5, we will return to discussions of resistances and oppositions. For now, our focus will be on the inscription of power on the landscape.

Multiple sites of power can be identified. For Karl Marx, best known for his iconoclastic work, *Das Kapital* (1957), capital and power were inseparable and power is defined in terms of control over the means of production. But power is not only tied to economy. Foucault argued that disciplinary power, exerted at the level of the human body, affects the individual's ability to act. Power is also believed to reside in the state, in religious systems and in racialised and gendered ideologies, socially constructed. In this chapter, our focus is on the power of the state, capital, religious systems and racial ideologies. While we deal with them independently in the first part of the chapter, in later sections we emphasise the intersections of various ideological and power systems and the mutually reinforcing effects of such intersections. For example, we examine intersections of capital and race, state and capital and state and religion – and the resultant landscapes of power. At appropriate points, we illustrate such intersections using multiple scales of space: the nation, the city, the region, as well as the space of the object, building and even human being.

4.2 The power of the state: landscapes of 'nationhood'

States often have direct control over landscapes, using planning laws and other legal and fiscal devices. Sometimes, authoritarian regimes simply rule by fiat. Yet, in order to be effective in the purveyance of particular ideologies, landscapes are often put to hegemonic use. In what follows, we offer illustrations of authoritative authorship of landscapes in a colonial context, followed by examples of symbolic manipulations of landscapes in two different contexts: one in which landscape inscriptions can be made afresh and another in which negotiations with existing landscapes are necessary in the exercise of hegemonic power.

The power of the colonial state is nowhere more apparent than in the shaping of colonial cities, which the foremost scholar Anthony King (1976) defines as non-western cities resulting from contact with western industrial colonialism. Such cities are shaped by cultural contact, levels of social, economic and technological development and the power structure of colonialism. In particular, the power structure of colonialism was reinforced by the creation of the segregated city with a colonial sector and an indigenous sector for economic, social, political and racial reasons. This was achieved through legal means or implicitly with residential areas marked by cultural and economic deterrents. The segregation of areas was designed to achieve particular ends, the first of which was to minimise contact between colonial and colonised populations. For the colonial state, segregation acted as an instrument of control, both of those outside as well as those within their boundaries. It helped the colonial community to maintain its self-identity as 'master', thought to be essential if it were to perform its role. At the same time, segregation of the indigenous population made for easier control of 'native affairs'. Economically, it was also useful as it cut down the total area that had to be maintained and developed (the colonial quarters). Furthermore, it helped to preserve the existing social structure.

Singapore's landscape under colonial rule exemplified these ideologies (see Perry *et al.*, 1997). To regulate the appropriation of land for specific purposes, Sir Stamford Raffles, founder of modern Singapore, instructed the town committee to focus on remodelling the town according to principles which would facilitate public administration and maximise mercantile interest, inscribe public order in space and also cater for the accommodation of the principal races in separate quarters.

Central to Raffles' plan was an expansive central space on the north bank of the Singapore River devoted solely to public purposes and dominated by grand edifices such as a church, government offices and a court house opening out to a central square. These colonial structures, which epitomised the ideals of British governance, were flanked on the east by an equally expansive 'European Town', carved out as a residential area for the European administrative and mercantile community. Raffles also ordered that the swampy south bank of the Singapore River, hitherto occupied by Chinese traders and raft houses, be drained to make way for a line of wharf and warehouses along the bank. In time, this became the principal commercial heart of the town. In planning his new town, Raffles stressed that the mercantile community should have first priority in claiming advantageous sites.

In the laying out of public spaces, Raffles emphasised the importance of open, orderly arrangement, uniformity and regularity. In essence, a gridiron system of streets with separating rectangular plots formed the basis of Raffles' plan. Not only did the gridiron provide an equitable method of dividing the land in a new city formed by colonisation, for the colonist unfamiliar with the lie of the land, it was a means of simplifying spatial order in order to provide for a swift and rough division of territory (Mumford, 1961:224). Raffles was keenly conscious that careful allocation of land was crucial to the orderly growth and prosperity of his new city.

In planning for the accommodation of the increasing numbers of Asian immigrants, Raffles demarcated the town into 'divisions' or *kampung* for particular racial and occupational groups. The residential spaces of different racial groups were to be as far as possible segregated: the European Town was allotted the expanse to the east of the government reserved area, while the Chinese, Bugis, Arabs, Chulias and Malays were relegated to more peripheral locations in well-demarcated *kampungs*. Each *kampung* was placed under the immediate superintendence of its own chiefs or *kapitans*. These chiefs were then made responsible to the Resident for policing their respective jurisdictions (Buckley, 1984:57).

In effect, Raffles' town plan both specified the spatial configuration of the town's urban development and served to address vital questions underlying the founding of a colonial city such as those of establishing public administration and order, strengthening the economic base and accommodating a rapidly expanding and highly diverse population. Landscape and its structuring was therefore integral to the colonial enterprise.

Unlike colonial cities which were subject to open and outright control, in contemporary non-colonial and post-colonial contexts, different symbolic strategies have been adopted in the exercise and display of power. We focus first on the example of Canberra, built as a wholly new capital of Australia in 1908 when 'the population comprised just 1700 people and a quarter of a million sheep' (Saunders, 1984:52) and saturated with symbolic meaning that the Australian government wished to invest and convey. The role of the government in the construction of Canberra is perhaps captured in Saunders' (1984:32) description of it as a 'company town, and that company is the Federal Government'.

The name 'Canberra' is an Aboriginal word for 'meeting place' and captures the essence of the unity desired for a new 'nation'. This was most

Figure 4.1 Australia's capital, Canberra, originally planned by Burley Griffin for 25,000 people

evident in the design of the capital, originally planned by Burley Griffin for 25,000 people (Day, 1986:16), with its sacred axes and the symbolism of public buildings (see Figure 4.1). In the location, architecture and design of public buildings, Canberra speaks of the role of the state in attempting to define an Australian 'nation'. Parliament House, opened in 1988, for example, is located on Capital Hill, with a roof cover of grass, so that the shape of the natural landscape was preserved.

It is crowned by an 81-metre flagpole, from which flies a huge Australian flag. Streets, named after the various state capitals, radiate from Capital Hill in every direction. Seen in another way, 'the whole country comes together on Capital Hill' and Parliament House is 'the navel of the nation' (Colombijn, 1998:572). Within Parliament House, further attempts at 'incorporating the nation' are evident: meeting rooms are coloured ochre after the Australian outbacks and grey-green after the eucalyptus forests and different woods typical of the Australian states are used (Colombijn, 1998:572). This represents part of the desire to legitimise the government by emphasising its oneness with the land, drawing on the image of the indigenous people as being 'with the land' (Hamilton, 1990:22; see also Box 2.2). Outside, at the other end of the main axis of the city from Parliament House, is the Australian War Memorial, decorated in such a way as to remind one of a church. It, however, does not make explicit references to Christianity at all, thus conveying the symbolic message that allegiance to the Australian nation constitutes a civil religion which entails the noble act of sacrifice. Its intent is the conveyance of the Australian national as 'tough, resourceful, independent and comradely'

(Colombijn, 1998:579). Both Parliament House and the Australian War Memorial draw symbolic strength not only from their architecture and design, but also from their location along the main axis of the city.

Along this main axis between Parliament House and the Australian War Memorial is the High Court building, which is suggestive of other symbolic nuances. The powers of the legislature and the judiciary are carefully balanced here, a response to the Chief Justice Sir Garfield Barwick's view that the Court building should 'express in form and place its function as a check on the legislature' (Colombijn, 1998:576). The outcome was a Court House neither too close to Parliament House nor visually dominated by it. It had a clear view of Parliament House, made possible in part by a glass wall on the appropriate side of the building (Colombijn, 1998:576). Canberra's existence and its spatial structure and architecture are therefore the outcome of very explicit powers of the state in 'writing' the landscape. Such symbolic capital is possible precisely because the city was a newly planned one and positional meanings could be inscribed on land previously unbuilt. (See Box 4.1 for an examination of another symbolic city, Venice.)

The case of Canberra is significant as an example of a newly planned city and may be revealing of similar experiences in other such cities (for example, Brasilia). It is evidence of the state using the landscape to naturalise and legitimise its power, not through direct force, but the persuasive symbolism of landscape meanings. Clearly, here is hegemonic power at work.

Box 4.1   Venice, symbolic city

The city of Venice has been mythologised as a utopia, beautiful and harmonious, 'a perfect union or conjunction of society and place' (Cosgrove, 1982:145). Venice, in medieval times, was a major trading centre at the junction between the Holy Roman Empire and the Byzantine Empire. The city also held a special religious status, as the recipient of the body of Saint Mark the Evangelist in 828 AD.

In part, its geographical location was seen as giving rise to its perfection and harmony, as a fulcrum between west and east and at the junction of land and sea. The encircling sea was seen as a protection, which allowed the city to trade freely and successfully. By the sixteenth century, writers were representing the city as a pure maiden, protected by the sea from barbarian attacks. Cosgrove commented that 'Virginal Venice, surrounded by the sea, protected by its saint, could be imagined as a sacred eternal city' (Cosgrove, 1982:147). The representation is a sexual image and embodied representations of the city are always female in form while textual representations are always romantic and sensual (Cosgrove, 1982:162–4).

The city was not only at a balancing point of land and sea, but also possessed internal qualities of balance and harmony, which were deliberately constructed into the landscape throughout the sixteenth century. The three

(continued)

(continued)

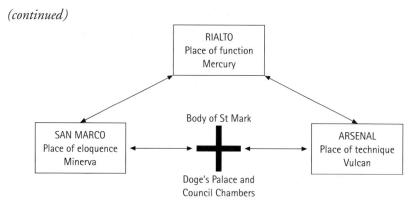

Figure 4.2 The three main functional areas of Venice: government, trade and exchange, manufacturing and industry
Source: Cosgrove, 1982:151.

main functional areas of the city held one another in counterbalance. The three areas consisted of San Marco, the seat of government; the Rialto, the locus of trade and exchange; and the Arsenal, the place of manufacturing and industry. These three areas are represented in Figure 4.2. Interestingly, this tripartite structure was replicated in the design of Canberra (see earlier in this chapter), with government located at Parliament Hill, commerce at Civic and industry at the third apex of the triangle. The industry never really developed in Canberra and this third point has subsequently become the focus of the defence industry. Nonetheless, the classic design of these cities reinforces hegemonic notions of city status and functions.

The ideology of balanced and harmonious perfection may be examined in more detail in Venice's San Marco, where the counterbalancing power structures are built around the Piazza (square) along a number of cross-cutting axes. In simple terms, the city represented not only a natural harmony between land and sea, but a monarchic power balanced both by a tradition of republicanism and by an influential aristocracy; a secular power matched by the city's sacred status; and absolute power tempered with wisdom (Cosgrove, 1982:149–51). All these features were deliberately inscribed in the landscape not only by fine architectural forms but also by the conscious positioning of the buildings which represented the counterbalancing forces.

The example of Venice is a complex and multi-layered one. The city itself has a symbolic structure at a number of levels. The myth of purity embodied in the city has not only been eulogised but has also affected the ways in which other cities have been constructed and texts written. The myth of Venice has repercussions down to the banalities of television advertising of romantic interludes on the canals. It is in such ways that the construction of power in the landscape comes to permeate the everyday and to have continual reper-cussions in our cities, our books and our lives.

While monuments and morphology can be used to assert authority and presence or to negotiate power relations quite comfortably in a new city such as Canberra and Brasília, in many other places, the state must contend with existing landscapes. Indonesia's Jakarta, and in particular, Medan Merdeka, a centrally located square in downtown Jakarta, offers abundant grist for the analytical mill here. In the nineteenth century, when the Dutch were still ruling in what was then known as Batavia, the square was the site of the governor's residence, police headquarters, a Roman Catholic cathedral, the Prince Frederick Castle and a host of government buildings. It was known then as Koningsplein (King's Square). It was clearly the centre of Dutch power. When the Indonesian government took over at independence, the square was transformed, reflecting the transference of power. Interestingly, there was no outright rejection of Dutch or, more accurately, European order and the notion of statehood – certainly, the square was not simply demolished. This

> appropriation and realignment of the space previously dedicated to Dutch subjugation of the colony is . . . consistent with an indigenous conception of power relations in which the power of vanquished enemies is absorbed to augment the potency of the victor (Anderson, 1972). *Medan Merdeka* is a more powerful symbol of independence precisely because it invokes the centuries of Dutch rule. It is no surprise then that the nationalists would reinforce the legitimacy of the new state by assimilating the spatial symbols of Dutch authority. However revolutionary the political rhetoric of Indonesia's leaders, the fact remains that their claim to the whole of the territory of Indonesia ultimately rested on acquiring it from the Dutch (MacDonald, 1995:278–79).

Several features of Medan Merdeka will illustrate the power relations just explicated. One of the remaining structures from Dutch colonial times is a neo-Gothic Dutch Cathedral. Its continued presence in such a prominent position, when Roman Catholicism is the religion of such a minority, carries the significance of the newly independent government's desire to weaken the significance and offset the presence of Islam. This reflects the view that, while Islam is an important religion in Indonesia, it is not an Islamic state. This symbolic gesture is, however, balanced by the fact that an imposing Istiqlal Mosque also stands on the square, southeast Asia's largest and the second largest in the world. The mosque stands in the place of the former Prince Frederick Castle, reflective of the displacement of European presence. And yet, as part of the balance of power, MacDonald (1995:286–7) argues that its Arabic dome, as opposed to a southeast Asian pyramid-like pavilion, symbolises the government's desire to represent Islam as an outside influence, like European colonialism, which has been incorporated into Indonesia's sociocultural fabric. Other new monuments that have been constructed at Medan Merdeka further symbolise the power of the new Indonesian government. One prime example is Monas (*Monumen Nasional*, or National Monument), an obelisk topped by a tall flame, symbolising the spirit of struggle in Indonesia's war of independence. As a towering edifice, taller than any Dutch construction in southeast Asia, Monas is Sukarno's tangible reminder of his inheritance of the former Dutch territories and his authority over the new state (MacDonald, 1995).

While monuments, by virtue of their size, are prominent landscape features and invite attention and analysis, other less imposing elements of the landscape are nevertheless as important for present purposes. Street names, for example, are highly visible aspects of our landscapes although they do not have the overwhelming presence of monuments and edifices. Their everyday use and functionality also mask the structures of power and legitimacy that underlie their construction and use. They map space and time in particular ways, reflecting a politics of landscape. We will pay attention in this chapter to their role in expressing, reifying and reinforcing power through hegemony and return to the ways in which they reflect conflict and resistance in Chapter 5.

In various ways, street names are intimately and integrally related to power relations. The act of naming streets is a way by which official agencies appropriate public space, possibly for political ends and because the contemporary norm is for the official naming of streets to be carried out by the state's agencies, the administrative procedure becomes an expression of state power. For instance, street names can naturalise a hegemonic version of history, often designed to consolidate political regimes. Such histories celebrate official constructions of the past as definitive and ignore, indeed, actively deny, other versions. When street names that commemorate particular (official, hegemonic) versions of the past (through celebrating particular events and personalities) are introduced, they serve to concretise and naturalise these privileged, constructed versions of the past by linking the past and its myths with the landscape (Azaryahu, 1996:320). In addition to drawing on historical resources, official agencies may also express their power through inscribing desired ideologies on the landscape via street naming. Often, these desired ideologies form part of the arsenal of 'nation' and national identity construction. By their everyday nature and their recurrent use, familiar street names serve to reaffirm the validity of a particular version of history and/or a particular construct of the 'nation'.

Multiple examples in varied contexts may be cited to illustrate these arguments: reflecting diverse ideological assertions, including communist, revolutionary, royalist and post-colonial situations. The French Revolution, for instance, ignited the practice of political representation through the renaming of streets and squares, evident in the demolition of the statue of Louis XV in 1792 in the Place Louis XV and the erection in the following year of the figure of Liberty and the renaming of the square as Place de la Revolution (today known as Place de la Concorde). The new state of Singapore, on gaining independence in 1965, sought to sever colonial apron strings and affirm a sense of local identity. Ground rules established in 1967 sought to steer clear of 'old colonial nuances, British snob names, towns and royalty' (Kong and Yeoh, 2002) and give priority to local names – new powers, new allegiances, new street names, new landscape meanings came hand in hand. Yet, local names had to reflect the political ideal of multiracialism and so there was to be a conscious departure from previous practice of using Malay names for roads, especially the use of the Malay word 'jalan' for 'road', unless used in tandem with other languages. Thus, in the Jurong Industrial Estate, streets were variously named Fan Yoong Road (meaning 'prosperity' in Mandarin), Jalan Tukan (meaning 'skilled craftsman' in Malay), and Neythal Road (meaning 'to weave' in Tamil), a deliberate and self-conscious effort at upholding the dominant ideology of multiracialism and its concomitant multilingualism.

In addition to states and nations seeking to invest their meanings in landscapes via naming practices, the manifestation of power can also involve reference to significant characters and personalities to gain political capital. In 1964, for example, Kirya, the Israeli government district in west Jerusalem named its streets after the patriarchs, judges, kings and prophets of the Old Testament (the emphatic maleness reaffirming the masculinity of public space), as a way of insisting on a direct symbolic link between the modern state of Israel and ancient Jewish history. This was an act of seeking legitimacy by naturalising Judaism through the landscape. In the former Union of Soviet Socialist Republics (USSR), a different commemoration process was at work. Soviet leaders were prominently commemorated, not only in street names, but in ships, hospitals, schools and towns as well. This became popular after Lenin's death in 1924, and grew disproportionately in the personality cult of Stalin in the 1930s and 1940s (Azaryahu, 1996:315). However, as recent events in that part of the world reveal, not least the unravelling of the USSR, the forced assertion of ideological constructions on the landscape can quickly become dismantled. The destruction of monuments and statues and the changing of street names demonstrate this. The effects of ideological impositions are not as strong as when hegemonically effected.

4.3 Capital, class and the rewriting of landscapes _____

Marx's concept of power, as highlighted earlier, is integrally tied up with capital. Capital rewrites landscapes in remarkably conspicuous ways, although other ideologies and powers also simultaneously seek to make their mark. Thus, towering skyscrapers dominate skylines and speak of the power of capital to saturate landscapes with their ideologies and meanings but also reflect the design profession and the conflicts between the two. In this, New York's commercial landscape of the latter half of the nineteenth century lends itself to scrutiny. The city's emerging mercantile and entrepreneurial class were chiefly responsible for commissioning New York's skyscrapers at that time. They sought to communicate power and prestige to potential customers, calling for prominent structures that would symbolically shout out their new wealth and, at the same time, serve as a form of advertising. Further, they were to bestow cultural legitimacy on those who commissioned them (Domosh, 1989:34). Citing the historian Frederic Cople Jaher, Domosh (1989:34) argued why New Yorkers, more than Bostonians, commissioned buildings that were more ornate and towering. The New Yorkers were 'more inventive, their firms had shorter lives, and they were greater credit risks. Because they lacked stability and cohesiveness, New York's commercial elites wanted physical expression of their power' (Domosh, 1989:34). This was particularly true in the case of the highly competitive, relatively new industries of newspaper publishing and life insurance, which relied on reaching out widely to an urban audience. Many of these were also owned and run by magnates who were keen not only to advertise, but to assert their corporate egos and to affirm their cultural legitimacy as arbiters of art and good taste. They therefore commissioned tall, imposing structures and sought to insert their own visions of aesthetics and cultural value. One example of the assertion of power, prestige and aesthetic judgement is the building for the newspaper, the *New York World*, completed in 1889, designed by George Post for Joseph Pulitzer. Domosh (1989:35–6) described it thus:

> The sixteen-storey building . . . towered six stories above the tallest buildings in the city. The gold dome clearly stated Pulitzer's desire for his newspaper to reign supreme in the world of New York journalism. Pulitzer himself added the dome as well as the three-storey arched entranceway to Post's design . . . When the building opened, a writer for the *New York World* stated that Mr. Pulitzer had insisted 'the structure must be in every sense an architectural ornament to the metropolis; that it must be a magnificent business structure of the first order, embodying the very latest and best ideas in construction; that, to be worthy of the paper it housed, it must be the best equipped newspaper edifice in existence'.

Yet, the power of capital did not always confer cultural legitimacy, for the kind of conspicuous building that merchants desired compromised the architectural beauty in the eyes of designers. It was argued that buildings which were constructed mainly for profit (usually in haste) did not result in a work of art. Further, the ornamentation that clients desired in order to achieve prominence ran against the grain of contemporary aesthetic evaluations (Domosh, 1989:36–7).

The towering skyscrapers that Domosh writes about are not the only manifestations of the power of capital. As Cuthbert (1995:298) pointed out, the near totalising power of capital is also reflected in the:

> sterilization of urban space by removing any historical referents, building on landfill and in the purging of nature; the simplification of urban form through lease agreements standard details and the use of a standard building prototype (the pavilion with several floors of parking under) and the sanitization of social space through corporate architectural ideology.

In many parts of Central Hong Kong, much of this has been achieved through monopolistic practices, with spaces made over and dominated in large areas of the city by a few economic interests, for example, the New World Development Group in the central area of Kowloon, Tsim Sha Tsui; Hopewell Corporation in Causeway Bay; and Hong Kong Land, a subsidiary of Jardine Mathieson, in Central Hong Kong. Their domination is reflected in their commodification of urban space and form, which in turn commodify social relations.

Cuthbert (1995; Cuthbert and McKinnell, 1997) discussed the commodification of public urban space using his concept of 'ambiguous spaces', those seemingly public spaces which are nevertheless owned and subject to control and surveillance by corporate powers. Increasingly, he argued, control of social space has been transferred from the public to the private sector in three ways. First, large new building complexes belonging to banks, insurance and property companies, multinational corporations and the like, are encouraged to create and donate 'public space' at ground or podium level and, in turn, are rewarded with plot ratio benefits. Ironically, however, once constructed, these spaces are given back to private ownership, in that use of the space is controlled privately. Second, pedestrian movement is channelled through corporate space. Third, large shopping centres are encouraged to provide open space, such as internal atriums and courtyards, which 'replace civic space with the commodity space of the market' (Cuthbert and McKinnell, 1997:296). These spaces 'masquerade' as social space but are actually commodified and controlled space.

Because of the continued private ownership of such urban spaces, they are open to surveillance by corporate powers and this occurs through physical policing and technological surveillance (electronic means, e.g. video cameras). The case of Jardine House in Hong Kong will illustrate the point. Jardine House appears to be surrounded by public space in the form of landscaped gardens, fountains and open space. Yet this space is under Jardine's control, evident when it decided to stop Filipina domestic workers from using these spaces as a meeting place on their one day off work a week (see the discussion of Little Manila in Chapter 3). The area was taped up and employees were tasked to police the area and remove 'offenders'. Video cameras were also installed for surveillance purposes (Cuthbert and McKinnell, 1997:300–301). The right to occupy certain open spaces in the city is thus whittled away by the power of capital to control its use, made possible in the first place by the state which granted the domination of such space to corporate power.

The contemporary shopping mall is also an unmistakable manifestation of the power of capital to reshape landscapes. Shopping malls developed as consumer spaces from the 1960s in association with the rise of the suburbs and the private car. Many of them are ordinary places, developments of the old main or high street where shopping is a little more packaged and a little more associated with food and entertainment than formerly. However, the major malls are fantastic places and are landscapes and employers of size and significance. Hopkins (1990:5) described the breathtaking size and scale of Canada's West Edmonton Mall, which alone takes one per cent of Canada's retail sales, employs the equivalent of a small town's entire population (18,000 people), and extends over a landscape of 110 acres, including an 18-hole golf course, a seven-acre water park and a replica of Miami Beach.

The construction of shopping malls of this kind reflects the reification of a number of myths (see Chapter 2 for a discussion of the myths of place). Perhaps one of the best ways of illustrating this is to examine how developers create a place image for shopping malls and how different landscapes are

Box 4.2   Fantasy: elsewhereness to placelessness

Shoppers enjoy the experience of shopping malls, not just to purchase items but also to enjoy the experience of being transported to another place and time. This can be a minor sensation such as the twinkling lights of fairyland or can be a major feature such as the ability to escape to sun, surf and 30 degree Celsius at 'Miami Beach' while being in the chill reality of a Canadian winter.

Shopping malls often have elements of simulated elsewhereness. Some of West Edmonton Mall's advertising tells us:

> Tourists will no longer have to travel to Disneyland, Miami Beach, The Epcot Park, . . . Bourbon Street, New Orleans, Rodeo Drive, Pebble Beach Golf Course, California Sea World, The San Diego Zoo, the Grand Canyon White Water Rafting. It's all here at the West Edmonton Mall. Everything you've wanted in a lifetime and more (quoted in Hopkins, 1990:12).

Hopkins argues that consumers are willing participants in this fantasy which they recognise as being different from the real thing. Consumers are, to some extent, voyeurs, gazing on difference. Furthermore, Hopkins (1990:8–9) argues that shopping malls use much more than a static representation of elsewhere, but construct consumable sanitised environments which have historical, natural or exotic referents as a spatial strategy to induce the consumer to stay longer and spend more. West Edmonton Mall's strategy of icons and elsewhereness helps differentiate this shopping landscape from the chain stores that can be found almost anywhere else. However, its pastiche of elsewhere renders it to some extent not just a place apart, but also placeless. The simulation of this strategy by numerous other mega-malls has led to the charge of a postmodern blandscape – a landscape that could be anywhere and which has so many referents to other places that it has gone beyond elsewhereness to placelessness.

---

appropriated and reinvented or reinvested with meaning in these malls to achieve their goals (see Box 4.2). Among the most stark must be the ways in which nature – perhaps that which is most separate(d) from humans – is exploited and commodified, exemplifying the power of capital over nature. Such exploitation and commodification is undertaken to achieve various ends: to 'soothe tired shoppers, enhance the sense of a natural outdoor setting, create exotic contexts for the commodity, imply freshness and cleanliness, and promote a sense of establishment' (Goss, 1993:36). In this presence of nature, consumption is naturalised, in the hope of mitigating 'the alienation inherent in commodity production and consumption' (Goss, 1993:36).

Several strategies of commodification are apparent: at the Mall of America, for example, nature is commodified in Rainforest Café ('an enchanted place for fun far away, that's just beyond your doorstep' (Goss, 1999:54)), Camp Snoopy and the Underwater World. There are also animals for petting in shops such as Nature's Wonders, Wilderness Station and Wilderness Theater, while names of stores also evoke 'pristine and mysterious nature', such as Forever Green, Natural Wonders and Rhythms of the Earth (Goss, 1999:60). An 'essentialist Minnesota sense of place' is recreated through a stretch of northwoods stream, 30,000 plants and trees (the largest in indoor planting in the world) and an artificial 70-feet waterfall and plaster cliffs cast from the originals along the St Croix River, complete with animatronics moose and bird noises (Goss, 1999:50). By naturalising consumption, particularly of goods and services that are not essential, the effect of capital is to legitimise the acts of shopping, purchasing and consuming and to reinforce the ideology of consumerism.

The shopping mall seeks to effect the power of capital over consumers in yet other ways, such as the construction of the Paradise myth. All shopping malls contain elements of the Paradise myth, even if it is merely the use of palm trees, hanging baskets or water features to induce a feeling of calm and well-being. The control over the extremes of climate, the background music and the removal of the outside world are conducive to a pleasant experience, which have some clear referents to Utopia or Paradise.

The outside of many shopping malls is bleak and bunker like, the buildings exhibiting tall blank walls to the outside and flanked by serried ranks of car parks bleached in sunlight. The unappealing outside creates a deliberate contrast to the pleasures within, where music, foliage, columns and arches combine to provide a shady refuge where climate is controlled, noise is hushed and odours sanitised. In this pleasant interior, however, the spoils still have to be striven for and there is an element of tension and competition created, even if it is only by the mall furniture creating obstacles to rapid progress (and thereby keeping the consumer there longer). Nonetheless, the goods are laid out in enormous bounty, piled high (often on hidden sloping boards) to give the overwhelming impression of a Paradise overflowing with plenty.

In an article about Australian shopping malls, Michael Duffy described them as 'Australia's new cathedrals'. He argued that shopping malls are changing and enriching the suburbs, providing 'air-conditioned and picturesque surroundings' (Duffy, 1994:28) which make shopping a leisure-time activity to be enjoyed rather than endured. Although he does not labour the point, the implication of his statement about the new cathedrals is that they are providing the focal point and the spiritual refreshment that churches used to provide within communities. A specific example of such analysis is evident in how the gateway to the Wollongong Mall in New South Wales, Australia, casts the shopper as pilgrim, the mall as Paradise. The massive blank walls of the outside of the mall remind one of a fortress which must be scaled if the shopper is to obtain the 'promised land' and draw on the idea that good results come only with struggle (Winchester, 1992:146–7). The good(s) that resides once inside are in 'lavish piles', offering a 'veritable cornucopia, a vision of bounty' (Winchester, 1992:146). Myths about Paradise and the promised land are therefore embedded in the landscape, naturalising the desire, the search and the promise of paradise, to be realised in the 'lavish pile'.

The mall is not only an earthly shopper's Paradise but it is also represented as a consumer Utopia. A mall removes the shopper from the everyday and the banal, from the pressures of clocks and weather. A mall goes further in encouraging consumers to believe that they can be personally transformed or enriched by their experience (as indicated by Duffy, 1994) and, moreover, that that transformation has no bounds in reality but is a choice available to everybody. Average people are encouraged to believe that they can do anything they want; see the world, tame nature or follow the sun. In this way the mall embodies myths of escape, egalitarianism, personal transformation and millionaire status, while disguising the workaday and the social relations of production.

The mall is also encouraged to develop its profile as a public congregation centre with a number of community events, such as educational fairs and community group performances. In Australia, malls are even seen to be re-creating marketplaces (Duffy, 1994). However, there is a major difference in that malls are essentially private and controlled spaces, where non-conforming users (such as buskers, beggars and youth groups) may be identified, supervised and, if necessary, ejected (Winchester, 1992).

The layers of meaning in the landscape of the shopping mall diffuse, legitimise and transform the consumption process. The scale of the mega-mall is utterly dominating; increasingly its architecture is spectacular. References to Paradise and elsewhere remove shoppers from the immediate social relations of production and imbue them with an ethos of personal transformation, choice and fulfilment. Recent estimates have indicated that in Canada time spent in shopping centres ranks third after home and either work or school for the average Canadian (Anand, 1987 quoted in Hopkins, 1990:7). The landscape of the shopping mall is therefore a peculiarly significant social control over people's lives and one which is imbued with layers of meaning which legitimise and naturalise personal consumption as a valuable activity.

Yet other strategies of creating 'a world apart' into which shoppers are tempted to dwell are evident in the landscape. For example, there is the deliberate design element where windows are largely lacking in shopping malls to prevent shoppers from gaining a glimpse of the outside world (Goss, 1993:32). This has a dual purpose: to create a feeling of being in an alternative world and to disallow any possibility of telling the time. The effect intended is that of a timeless Paradise of commodities. Shoppers in Tangs, a major shopping centre in Singapore's Orchard Road, have shared their sense of timelessness, talking about 'losing track of time', of not realising it was pouring outside, of forgetting their other appointments because they were in the 'lost world of shopping' and so forth. The strategy of shutting out the world outside and creating a timeless landscape of Paradise within is further encouraged by other elements of the landscape which urge shoppers to look within and explore. For example, given the way in which many shopping centres have multiple floors, designers and management attempt to draw shoppers towards other floors apart from the one they are on. One way is to offer them glimpses of the displays on other floors. In a number of Singapore shopping centres, it has been observed that glass elevators are used; balloons, vertical banners, drapes of cloth/ribbon are introduced; and sculptures installed, which have the effect of drawing the eye vertically (Kong, 1995; see also Maitland, 1990). Shoppers may then be drawn simultaneously by the window displays on other floors.

The main participants of and targets of such naturalisation and legitimisation of consumerism through various means are women. Prime sites tend to focus on women's consumption goods and much less on the male market. This focus on women is significant in two ways. First, it reinforces the roles of women as wives, mothers and daughters as carers and givers, underlining the particular conception of womanhood and the ideology of motherhood, as it perpetuates their role as major purchases of goods for the family. Second, it underscores the perception of women as objects of sexuality, encouraging their beautification and adornment. At worst, it suggests that women are the decorative accessories to men who conduct the real business of the city outside.

The manipulation of nature and the reproduction of fortress/Paradise in mall landscapes is matched by the wholesale building of a fantasy land by corporate interests in the shape of places like Disneyland. Landscapes, in such cases, are thoroughly manoeuvred for market use. Like nature, which is manipulated to enhance enjoyment, theme parks like Disneyland and EPCOT create the perfect worlds of the past, present and future for display. As Zukin (1991) pointed out, EPCOT's Paris does not reveal the reality of urban decay, neither is the overcrowding in China evident in the EPCOT version. The social and moral problems epitomised in the brothel are also absent from Main Street USA. 'Nostalgic and non-threatening fantasy images . . . mask the true economic, social, and political relationships that underlie the Disney Corporation', offering no hint that it is the 'product of a powerful, multinational corporation acting as a virtual totalitarian state and shrewdly advertising a global marketplace of commodities' (Warren, 1996:554). Disney World, Zukin argued, is not only a fantasy land, but is symbolic of the possibilities of the future. As we illustrate later, state and capital collaborate to expand this fantasy world into the real world.

In many societies, ownership of capital is central to the construction of class (although in some instances, such as the Hindu caste system, this may not necessarily be the case). As with capital, class structures cause (and are reinforced by) particular landscapes. Here, we illustrate with the example of gated communities as elitist landscapes.

Gated communities are one of the clearest landscape manifestations of the desire among the upper class to assert its presence and elite status through the creation of the image of elite landscapes. Gated communities may be lifestyle communities (e.g. country clubs) or residential communities with access controlled by physical barriers such as gates, fences and guarded entrances. These communities restrict public access not only to the residential buildings but also to the roads, pavements, playgrounds and other open spaces within the locality.

While some gated communities may be formed to keep out crime, more commonly, people of different classes set up gates for privacy, prestige and control over their homes, streets and neighbourhoods. The ability to do so through exclusion is an exercise of power or at least in seeking power through the creation of a 'hallmark of elite space' (Blakely and Snyder, 1998:61). That such exclusion often occurs in very low crime areas suggests that privacy and prestige value are frequently more important in motivating gating up. That gated communities now exist beyond the upper and upper-middle classes suggests that this 'once elite prerogative' is now extending to the middle classes and is provided by developers as a 'cachet of exclusive living' for those with 'less exclusive incomes' (Blakely and Snyder, 1998:63). Thus, while some gated communities are peopled by social and economic elites (thus constituting landscapes of power), most merely portray that image, bestowing a sign of elitism and a prestige status which social and economic class alone would not have afforded (thus reflecting the power of landscapes).

In this regard, Davis (1990) and others have argued that urban authorities have surrendered their reformist ideals of social integration. No longer do authorities and opinion makers act against social polarisation. Instead they are now adapting urban form, or transforming the morphology, to meet the requirements of social polarisation. For example, local governments in Australia have assisted with this trend allowing the establishment of gated communities, sometimes even subsidising the infrastructures of such residential estates (Hillier and McManus, 1994; see Box 4.3). In these respects, the urban landscapes of western cities are no longer emancipatory (Jameson, 1988:21–3).

Box 4.3   Looking within, on Sovereign Islands, Queensland, Australia

Gated communities typically have a wall, fence or moat surrounding them (Figure 4.3). These physical boundaries are purposefully deployed to alienate, exclude and differentiate (Hillier and McManus, 1994:92). These places turn people inward. They employ private security and detach themselves from any reliance on under-resourced public agencies. Strata committees develop and enforce their own rules of behaviour and responsibilities. The individual family groups turn inward again. The houses bristle with steel bars, cameras and warning signs.

The Sovereign Islands are located in southeast Queensland, just north of Surfers' Paradise. The Islands were created using dredged materials. They are in every sense socially constructed. A bridge links the mainland to this island of affluence. But those crossing the bridge must negotiate a checkpoint. And while on the island of affluence, you remain under the scrutiny of the watchtower and private security (Figure 4.4).

Members of gated communities endeavour to turn away people who are not like them. In Sovereign Islands this exclusion is done through the price mechanism of the housing market. Only the affluent can afford the A$800,000 entry price (at 1995). The properties are on quite modest sized plots, but the buyers are after something more than just modern comfort. The putting up

(continued)

(continued)

Figure 4.3 Looking after ourselves: private security within the gated community. Sanctuary Cove, Queensland, Australia

Figure 4.4 Crossing the bridge to Sovereign Islands, Queensland, Australia

of walls, and the building of islands, is mostly about differentiating the landscapes within and those without. It is fundamentally about using landscapes, and constructions of it, to demarcate between the people inside and those outside (Davis, 1990; Hillier and McManus, 1994:93–4).

In 1998, approximately 8 million Americans were thought to live in gated communities. The residents were mostly white, middle-class professionals. There are about 30,000 of these communities in the USA (Blakely and Snyder, 1998). There appears to be a considerable demand within western cities for these gated suburbs.

4.4 Race and the inscription of power

Quite as powerful as capital in the reshaping of landscapes is racial ideology, sometimes through the dominant allocation of meanings to places, and other times in the actual distributive use of space. The former is well illustrated in the racial ideology that shaped white policy towards Vancouver's Chinatown; the latter is manifestly apparent in South Africa's apartheid landscape.

Landscapes are ascribed meanings by different groups and often power gives the dominant groups a disproportionately important role in defining place meanings, oftentimes with material consequences. Anderson's (1987; 1988) research on Vancouver's Chinatown offers insights into how a racial category – white European – by virtue of its powerful, dominant position, defines the minority, less powerful racial category 'Chinese' and their place 'Chinatown'. She argues that both 'Chinese' and 'Chinatown' are not 'natural' categories but constructed within the white European cultural tradition (see also Chapter 2 on constructed hierarchies). By granting legitimacy to these constructions, Vancouver's municipal authorities in the late nineteenth and early twentieth centuries played the role of powerful agents in the definitional process.

Specifically, Anderson (1987; 1988) illustrated how the municipal authorities granted legitimacy to the ideas of 'Chinese' and 'Chinatown' by 'inscribing social definitions of identity and place in institutional practice and space' (Anderson, 1987:580). These definitions were invariably negative and in their 'social force and material effect' (Anderson, 1987:581), 'shaped and justified the practices of powerful institutions toward it and toward people of Chinese origin'. The negative definitions – 'their appearance, lack of Christian faith, opium and gambling addiction, their strange eating habits, and odd graveyard practices' – set the Chinese 'irrevocably apart' (Anderson, 1987:594). These characteristics led to the crystallisation of institutional practices – formal council designation of 'Chinatown' by medical health officers as an entity in itself deserving separate attention: alongside pig ranches and slaughterhouses, there should be checks by health officers on their rounds and in health committee reports of Chinatown. These spatial practices were deemed necessary because the Chinese 'method of living' was totally different from that of the white people – they could not be made to adopt sanitary methods and even where 'every convenience' was provided, the Chinese were 'generally dirtier than whites' (Anderson, 1987:587). It did not matter that the Chinese suffered, among other privations, job and pay discrimination not of their own doing, thus pulling down standards of living. Neither did it matter that the entrepreneurial sector of Chinatown attempted to change its physical conditions, not least through applying for its own licence to conduct street-cleaning operations, so dissatisfied was it with the City Government's substandard refuse collection in their area.

Attempts to curb landscape changes arising from the entrepreneurial spirit of the Chinese and their improving economic conditions point further to the power of whites to define people and place. In the late 1910s some ambitious Chinese merchants sought to move to the suburbs from their heathen 'cesspool' in Vancouver's Chinatown. Itinerant Chinese peddlers also brought their produce to their clients, thus leaving the contained boundaries of Chinatown and entrepreneurial Chinatown merchants began to open small stores with attached residences in locations more convenient to their clients. The response was swift and harsh: the 'yellow peril' had to be contained. As Anderson (1988:138) put it, the 'wartime "infiltration" of Chinese challenged the sociospatial order that was their [the European community's] hold on power and privilege'. This took various forms, for example, the municipal council sought to appease the (white) Vancouver Retail Merchants' Association by imposing a huge annual licensing fee on peddling. The council also sought to find ways of confining 'Asiatic retail business' to 'some well defined area' (council minutes, quoted in Anderson, 1988:139).

The spatial containment of the 'Chinese' as a racial group in Vancouver was writ large with far more widespread intent and consequence in apartheid South Africa. Williams (2000:167–8) summarised the racist planning frameworks of successive white-controlled governments by citing parts of a parliamentary speech by the then Minister of Native Affairs, Dr Hendrick F. Verwoerd in May 1952:

- Every town or city, especially industrial cities, must have a single corresponding black township.
- Townships must be large, and must be situated to allow for expansion without spilling over into another racial group area.
- Townships must be located an adequate distance from white areas.
- Black townships should be separated from white areas by an area of industrial sites where industries exist or are being planned.
- Townships should be within easy transport distance of the city, preferably by rail and not by road transport.
- All race group areas should be situated so as to allow access to the common industrial areas and the CBD without necessitating travel through the group area of another race.
- There should be suitable open buffer spaces around the black township, the breadth of which should depend on whether the border touches on densely or sparsely populated white areas.
- Townships should be a considerable distance from main and more particularly national roads, the use of which as local transportation routes should be discouraged.
- Existing wrongly situated areas should be moved.
- Everybody wants his [sic] servants and his labourers, but nobody wants to have a native location near his own suburb.

These racially driven planning frameworks created 'Islands of Spatial Affluence' in a 'Sea of Geographical Misery' (Williams, 2000:168).

Such pronouncements and actions were advocated on the supposed grounds of providing ' "fair and just" treatment of the African population' in that 'model' or 'garden' villages were being created to enhance the environmental and social conditions of the black population in the townships. Robinson (1997:369) cites the example of the New Brighton township known as McNamee Village in which the layout and housing design followed after the garden city. Amenities such as house lighting were provided as a way of 'improving' and 'uplifting' the African population. This was the direct outcome of views such as those held by Councillor Alf Schauder who uttered in 1949, that '[w]ithout adequate housing and associated ' "civilized" lifestyles, . . . a suitable "Native policy" was not really feasible' (Robinson, 1997:369). The firmly held assumption, as Anderson illustrated in the case of Chinese in Vancouver's Chinatown, was that African (and Coloured and Indian) people occupied 'filthy and dangerous' areas of town (Robinson, 1997:378). This served to justify the state's racially based intervention in urban management.

Both of these examples point to a (colonial) hierarchy of race, which we referred to in Chapter 2 (Boxes 2.1 and 2.2) and which we refer to again in Chapter 6. Legislative action reaching into various aspects of everyday lives, from housing to industry to schooling to transport and other aspects of urban management, stem from a deep-seated notion that there are 'better' races higher on the racial hierarchy and which deserve better treatment and more salubrious conditions than others. Landscapes and their preferred management naturalise this notion of the racial hierarchy (see Box 4.4).

Box 4.4 Aboriginal housing

European settlement on the Australian continent caused conflict over living environments. The colonial government, and subsequently, state and Commonwealth governments, attempted to address the conflicts by creating and maintaining reserves in which Aborigines were segregated, living under separate laws applicable only to them. In exerting their power to 'manage', 'protect' and 'advance' the 'welfare' of Aborigines, the housing landscape became a major instrument. Heppel (1979) highlighted how, in the 1950s, it was decided that Aboriginal people living on reserves would be provided with one, two or three-room dwellings without amenities as a first stage of 'management' and 'development', followed by progress to similar dwellings with basic amenities and then to standard, fully equipped suburban-type housing similar to those for non-Aborigines. This enforced segregation in low quality housing (for lack of funds prevented housing to develop to more standard suburban housing) reflected the integral role of the landscape in reinforcing government perceptions of and attitudes towards Aborigines, perceptions and attitudes which favoured their segregation and subordination. However, as Sanders (1993:215) pointed out, 'total segregation and protection of the Aboriginal population on reserves was never fully achieved, with some Aborigines continuing to camp informally on the fringes of settler communities, and others remaining beyond official supervision in remote localities'.

4.5 Gendered landscapes

At various points in this book, we have already alluded to the gendered nature of landscapes. Here we will give explicit attention to how gendered ideologies shape the landscape and how gendered landscapes in turn reinforce social constructions of gender roles and relations. In thinking about these issues, a useful tactic is to examine three scales of interaction, moving from a more macro to a more micro level. Gendered landscapes are first apparent in the separation of spheres, of males in the public domain and females in the private. At this level of social and spatial spheres and divisions, attention must also be paid to the dominance of heterosexual space, the outcome of the power of heterosexual ideologies and social constructions. Second, gendered landscapes must be understood in terms of the design of landscapes and the lack of attention to the needs of women. Third, at the level of individual elements of the landscape, such as the presence of public monuments and public imagery (for example, billboards and posters) and architectural style, gendered ideologies are also apparent.

Gendered landscapes were manifest in the 'divided city', a city separated into men's and women's spaces (Enjeu and Save, 1974; Loyd, 1975). Men's spaces were public and economic, anchored in production activities, while women's spaces were private and social, anchored in reproduction activities. These notions were reinforced by urban planning and design decisions (Wekerle *et al.*, 1980:8–9), most evident in the separation of the city into the centre and the suburbs. As Mackenzie (1989:114) summarised succinctly: 'Women's daily activities were carried out in opposition to a city made up of distinct work spaces and home spaces.' Yet, women worked in both spheres, engaging in both private and public spaces and activities, in production and reproduction. This demanded a dual role for women and systematically disadvantaged women because of the material constraints on their activities (Bowlby *et al.*, 1989:158).

Even while cities have been conceived and designed in gendered terms, separating between male and female spheres, the underlying assumption is undoubtedly that of a heterosexual city. Yet most heterosexuals are oblivious (conveniently) to the way in which space is often sexualised and of how it is in particular *heterosexualised*. Ordinary spaces – those of residence, work and activity space (thus, both public and private space) – are heterosexed (Chouinard and Grant, 1995; Kirby and Hay, 1997; Valentine, 1993a). In most societies, there is a heterosexual norm to which all people are expected to conform. Gays, lesbians, bisexuals and transgender people are portrayed as deviant from the norm, as the 'Other'. This heteronormativity is:

> institutionalised in marriage and in the law, tax and welfare systems, and is celebrated in public rituals such as weddings and christenings (Valentine, 1993a:396, 410).

Heterosexuals (unconsciously) claim spaces through the use of heterosexual signifiers such as pictures of partners and the use of hetero-centric dialogue. Valentine (1993a:410) outlined the political dominance of heterosexuality in ordinary English landscapes. In workplaces, private houses, restaurants and in public spaces, only opposite sex discussions or behaviours were tolerated. The workplace, too, is dominated by heterosexual performances, including opposite-sex flirting and heterosexual innuendo. Signifiers of heterosexuality include family photographs, wedding rings, posters and also hetero-centric gossip:

> In the job I'm in, they all talk about their men and their husbands and they've had a nice weekend and done this and done that. And I basically can't say anything (interviewee in her 40s, quoted in Valentine, 1993a:403).

Many lesbians felt ostracised from social events associated with work and generally felt unable to invite colleagues home for dinner parties. This was particularly a problem for those who were not 'out' at work.

Same-sex couples or families headed by a same-sex couple, must always be prepared for 'heterosexual incursions' into their space. Kirby and Hay (1997) reported how gay men in Australia felt compelled to 'straighten up the house' when relatives were expected. This straightening involved hiding the gay magazines and pretending that the spare room was the bedroom of one of them (Kirby and Hay, 1997:296–7). Restaurants and public houses are also places where heterosexual performances are overt. Same-sex couples sometimes experience hostility from staff and other customers. Diners find themselves steered into dark corners and to the poorest seats (Valentine, 1993a:403).

The extent of the heterosexual appropriation of landscapes and the assumed 'naturalness' of heterosexuality is such that the (hetero)sexed nature of landscape remains largely unremarked on. At the same time, the spaces and identities of gays, lesbians, bisexuals and transgender people remain pathologised. Non-heterosexual performances are deemed 'out of place' in most landscapes. Performances of sexuality are regulated through intimidation and violence:

> Heterosexual men and women can walk together safely in the streets; gay men and lesbians, in contrast, must negotiate the threat of violence each time they enter the public realm – particularly if they walk with a same-sex partner (Namaste, 1996:227).

There is a range of repressive outcomes and strategies that result from the spatial supremacy of heterosexuality. The political responses of gays and lesbians are thoroughly spatialised and include the construction of same-sex landscapes (see Chapters 5 and 6).

Apart from the separation of spheres (public and private) reflecting gendered ideologies and the hegemony of the heterosexual, a second way in which the power of male-dominated ideologies plays out in the landscape is in the ways in which landscape designs have been revealed as unfriendly to women in various ways. Underlying this body of work is an assumption that the design profession is male dominated and, as a result, women's needs are not considered.

Examples cited to prove how unfriendly landscapes are to women include, for example, the lack of ramps for baby pushcarts which women often have to handle and poorly lit, hidden corners which increase the chances of crime, including sexual crime against women. In the literatures on landscapes of fear, focusing on women's fears, in particular, various strategies of landscape planning have been encouraged, for example, the removal of environmental incivilities such as graffiti, vandalism and littering, which are associated with threatening environments; the provision of better lighting; the careful planning of landscape features such as trees, bushes and shrubs, which, while pleasant, can also block visibility, especially along pathways; the avoidance of blind corners and bends; and the introduction of mixed land use to ensure that there will often be people in different activities around for informal surveillance (Pain, 1991; Valentine, 1989; 1990; 1992).

A third illustration of power in landscapes, evidenced through male monopoly, is the landscapes of public monuments and imagery. Monuments reveal what we value and what we feel should be conserved and commemorated. Thus, by examining the monuments in our landscapes, we have a sense of a society's values. Monk (1992) and Warner (1985), for example, illustrate how few women are commemorated in public landscapes. This invisibility of women in landscapes of monuments, when challenged, only happens in one of two ways; first, within male spheres of politics or militarism, for example, when Queen Victoria is depicted as the symbol of empire and colonial power; and second, in abstract, impersonal and generic ways that embody noble ideals for men, for example, as nurses of soldiers during the war, who then become represented as mother figures tending the wounded; or as a symbol of liberty (the most famous being the Statue of Liberty, which cries out irony, given women's position in society through history). Not unlike public monuments which subordinate women in landscapes, public images of women are often objectifying. In posters, billboards and other forms of advertising, women are often portrayed as sex objects (see Figure 4.5). Women may tend to avoid places with such forms of degradation.

What we have illustrated in this section is that gendered ideologies shape the landscape while gendered landscapes also reinforce social constructions of gender roles and relations. The dominance of male ideologies has resulted in landscapes which privilege men and their needs and depict them positively, while women and their needs are marginalised and often depicted derogatorily. Similarly, heterosexuality is privileged and public spaces are heterosexualised so much so that homosexuals feel ostracised.

4.6 Intersecting powers

4.6.1 Capital and race

In apartheid South Africa, racial construction and forces of capital were often inseparable in contributing to the treatment of African labour. In attempting to achieve economic accumulation, segregation was effected through the

Figure 4.5   The public objectification of women, Parramatta Road, Sydney, Australia

confinement of African settlement to certain rural areas and encouragement of predominantly male migrant labour in order to secure cheap labour power (Wolpse, 1974; see, however, Simkins, 1981). In this sense, Posel (1988) argues that the reserve strategy – the patently geographical act of separating African labour in reserves – was aimed at fulfilling the interests of specific fractions of capital, namely agriculture and mining. Herein were to be found a ready concentration of African labour for white capitalists. At the same time, Posel (1987, cited in Robinson, 1997) also argued that these segregationist state policies could not be reduced to meeting the needs of capital only or to ensuring economic growth. They were as much about the preservation of white domination, the facilitation of surveillance capacities and the implementation of various other state policies such as the townships described earlier (Robinson, 1997). Landscapes and segregations therein therefore play a critical part in reinforcing the power of racial constructs and capital.

4.6.2 State and capital

Modern fairs, expositions and exhibitions are a nineteenth-century invention of social elites. Rydell (1984:2–3) argues that they represent the triumph of hegemony, 'confirming and extending the authority of the country's corporate, political, and scientific leadership'. It is difficult to tease apart and generalise about the relative power exerted and role played by the various segments of the elite. What *is* clear, however, is the critical role of financial institutions and corporations as guarantors of financial stability. Ley and Olds (1988:198) detailed, for example, the 'partisan nature' of sponsorship for various expositions: the 1939–40 New York World's Fair relied on sponsorship from 87 banking and trust companies, corporations, law firms, insurance companies, retail firms and business associations. In contrast, the three levels of government managed only 15 representatives, arts and education, eight, and organised labour, one. The role of the state, however, must not be diminished. It can arbitrate between competing claimants when it comes to the choice of city for an exposition. It can also override the powerful economic elites and cancel a fair, as evident in the 1992 Exposition in Chicago. Indeed, unlike the New York Fair, it can also contribute significant financial support: 76 per cent in the case of San Antonio's 1968 Fair; over 80 per cent of Spokane's 1974 Fair and 74 per cent of Knoxville's 1982 Fair (Ley and Olds, 1988:198–9).

As instruments of hegemonic power, expositions have economic, developmental and political goals in the form of developing trade, enabling local boosterism, enhancing political ambitions, consolidating political influence and offering education and entertainment. Funds procured through expositions have financed lasting legacies, from a new transportation system in Montreal (1967) and Vancouver (1986) to urban redevelopment (San Antonio, 1968; Brisbane, 1988) and cultural, sporting, recreational and convention facilities (Seattle, 1962). Trade and tourism have been enhanced. In political terms, expositions have celebrated the *entente cordiale* between two nations (e.g. Britain and France in 1908) and consolidated the legitimacy of leaders (e.g. Louis

Napoleon in 1855) (Ley and Olds, 1988:199). As the 1986 Vancouver World's Fair came to a close, the ruling political party called an election and was returned to power in British Columbia with the largest majority in its history (Ley and Olds, 1988:191). In many ways then, international expositions, as the product of elites, are rooted in the dominant ideologies of business and politicians.

Expositions were also educational in a variety of ways, selling ideas about 'the relations between nations, . . . the advancement of science, the form of cities, the nature of domestic life, the place of art in society' (Benedict, 1983:2). As entertainment, they were characterised by spectacle and grandeur, 'home to the superlative' (Ley and Olds, 1988:199), seeking to impress through their 'exaggerated proportions (both monuments and miniatures), their beauty, their exotic character, and their freakishness' (Ley and Olds, 1988:199).

The educational and entertainment value of expositions is, however, not necessarily innocent. Before 1914, for example, social Darwinism informed the ethnological exhibits, which reinforced racist attitudes (Ley and Olds, 1988:200). In exhibiting the pinnacles of scientific, intellectual, and aesthetic growth, national sovereignty was expressed, with the west affirming its superiority and its civilizing influence over colonies, while seeking to outdo one another within the 'civilized' world (Greenhalgh, 1988). Celik (1992) illustrates how Islamic architecture, cultural expressions and social life were displayed in the world's fairs in Paris and London in the nineteenth century – through orientalist lenses – as a decadent Near East, with harems, belly dancing and ululations which form part of the unknown and unknowable. Colonial peoples from south Asia, southeast Asia, the Americas and Africa were portrayed as having benefited from the civilizing process, made possible with colonialism (Greenhalgh, 1988). In the 1904 Louisiana Purchase Exposition, for example, the Philippines people were depicted as 'primitive' but on the verge of being Americanised.

In brief, it is 'rivalry between cities [that] takes the form of creating a high quality image through the organization of spectacular urban spaces (Harvey, 1990; Smith, 1990)' (Colombijn, 1998:565), resulting in, *inter alia*, urban mega-projects such as the world fairs just discussed. They result from the power of the state and capital to raise huge sums of money and handle massive logistics to make over entire landscapes for a significant number of weeks.

Other mega-projects also speak of this intersection of state and capital in the creation of place. One prominent example is the Disneyfication of landscapes, a process by which the principles of the Disney theme parks appear to dominate more and more sectors of society, not least in urban planning and landscaping, but also in its effects on social and economic relations. Such Disneyfication is perpetuated by the Disney Company itself, which pursues outside planning and design commissions (Warren, 1994). These projects appear to blur the boundaries of fantasy and reality. They also suggest that ultimately, the power to transform landscapes is not always effected positively or successfully, inviting resistance.

Sorkin (1992) illustrates this Disneyfication in terms of how the theme park has extended outwards beyond its confines and recast urban landscapes as 'variations on a theme park'. State planners and urban designers have called on corporate experience, seeking to replicate both the look and underlying structure of Disney theme parks (Blake, 1972; Boles, 1989). Thus, the principles and organisational techniques of the theme park (the 'imagineered logic', to use Relph's (1991:104) words) are being extended into urban areas, including 'perceptual control, centralised provision of goods [and] services, and conglomerate organization and ownership of attractions' (Davis, 1996:417). This has entailed planners 'seeing with "imperial eyes", that is, with little recognition of the existing human population or social history, except as potential labor and potential attraction, respectively' (Davis, 1996:417). Principles and practices of Disney theme parks have pervaded not only the planning and design of whole cities, specific elements of urbanscapes in the form of shopping malls, festival markets, small town main streets and residential neighbourhoods have also been 'co-opted by the mouse' (Warren, 1994:89).

The Disney model is attractive to city planners because it appears to offer solutions to urban problems and a 'powerful and comprehensive urban vision' (Warren, 1994:96). This is aided by the Disney Development Company, a subsidiary dedicated to applying the lessons Disney had learnt in Disneyland to urban development in all its various forms, encouraged by its CEO of the 1980s who exhorted cities to apply Disney's principles of design, crowd management, transportation and efficient entertainment to urban spaces. However, from another perspective, Disneyfication has been described as 'sinister' (Sorkin, 1992:xiv). Resistance against Disneyfication is evident, a matter to which we will return in Chapter 5, where we outline the specific experience of Seattle, illustrating the promise and difficulties of transplanting Disney World into the real world. For now, we examine the exercise of power by state and capital in the Disneyfication of Seattle.

Seattle Center was an aging civic centre area originally constructed for the 1962 World's Fair. Disney was engaged as urban planning consultant in order to inject the principles championed by Disney and, it was hoped, to enjoy its people-friendly and efficient designs. As part of its arrogance (and perceived power), Disney indicated that they wished to develop and possibly finance and operate the entire site, offering input in architecture, design, site layout, landscaping, crowd and traffic management and security. It would 'reshape the center, organize the chaos, and harmonize the currently inefficient use of space' (Warren, 1994:100). For sure, it did not wish simply to finetune the amusement zones. It nevertheless assured Seattle that something unique was to be created for the area and there would not be mere replication of Disneyland or EPCOT. Unfortunately, despite its claims, Disney seemed unable to create designs that took into account the needs and desires of Seattle residents and would neither seek nor take advice from locals. All three plans it submitted appeared dysfunctional and unappealing, did not consider Seattle's unique character and recycled the same ideas. Indeed, Disney proposed to demolish several cherished structures, replacing them with buildings and activities deemed inappropriate for Seattle's needs. The media, citizens' groups and council chambers alike grew increasingly dissatisfied and unhappy and, despite a renewed effort at a fourth plan, the damage had been done. The final straw was when Disney projected that it would cost US$335 million, a far cry from the original US$60 million estimation (see also Box 4.5). Disney was sacked, underscoring the fundamental contradiction that the charming fantasy and efficient infrastructure of Disney World could only be achieved with unacceptably authoritarian planning practices in real life. Local architects, planners, designers and other citizens were called in to do the job and the newly renovated Seattle Center became symbolic of the rejection of 'an autocratic, outside force in order to retain control over their space' (Warren, 1994:104). Once again, the imposition of an external vision and the attempt at direct control of the landscape, reflecting a lack of effort at persuasion and hegemony, has resulted in a rejection of the imposed vision and landscape.

4.6.3 'State' and religion

The conflation of religious and political power can be a potent combination. Such power is nowhere more evident than in the ways in which religious and political ideologies are mutually reinforcing in the making of landscapes. This could be the creation of a religious landscape that supports a political position, as the following example of Paris' Basilica of Sacré Coeur (Harvey, 1979) illustrates or the adoption of particular religiously inspired landscapes for political legitimacy as Duncan's (1990) study of Kandy show. Harvey (1979) illustrates this with the example of Paris' Basilica of Sacré Coeur.

The moral decadence of Napoleon III's Second Empire and the excesses and turmoil of the Franco-Prussian War of 1870–71 and the Paris Commune[1] (e.g. tensions between different factions of the ruling class; tensions between a traditional bourgeoisie and a working class; appalling conditions in Paris – including living off cats, dogs, rats and the zoo's elephant!)

> plunged Catholics into a phase of widespread soul-searching. The majority of them accepted the notion that France had sinned and this gave rise to manifestations of expiation and a movement of piety that was both mystical and spectacular. The intransigent and ultramontane Catholics unquestionably favored a return to 'law and order' and a political solution founded on respect for authority. And it was the monarchists, generally themselves intransigent Catholics, who held out the promise for that law and order. . . . There was little to stop the consolidation of the bond between monarchism and intransigent Catholicism (Harvey, 1979:374).

It was at this time that the Basilica of Sacré Coeur was being proposed and planned. The powerful alliance between the religious (Catholics) and the political (monarchists) caused its construction to be turned into a national project on the hill in Paris known as the Butte Montmarte. Here, some of the most significant opening and closing events of the Paris Commune had occurred, including some violent deaths of the Communards. The construction of the Sacré Coeur by the monarchy-supporting Catholics at the site where

Box 4.5 Darling Harbour: legitimising state investment

Darling Harbour, in the centre of downtown Sydney, was transformed in under five years from derelict railway yards to a prominent tourist and recreation project complete with monorail and (later) a casino. Darling Harbour was predominantly a state project, supplemented by private funding in the later stages, particularly for the casino and associated hotel development, which was initially controversial.

The Darling Harbour site, which in area accounts for over 40% of the Sydney CBD, had fallen into disuse as a result of technological change and industrial restructuring (Huxley and Kerkin, 1988). In the 1971 Sydney Strategy Plan the area had been designated as parkland, and later as an inner-city suburb, as a strategy to revitalise the inner city. Neither of these projected uses materialised.

In May 1984 New South Wales Premier, Neville Wran, announced a revitalisation scheme to include a high technology family entertainment discovery village; an aquarium; hotel/casino complex; exhibition centre; convention centre; harbourside festival retail markets; waterfront promenade; parks and gardens; and, was to incorporate the Maritime Museum and Power House Museum (Huxley and Kerkin, 1988). At the time of the announcement it was anticipated that the required funding of A$200 million would come entirely from state sources, but by the time detailed plans and costings were released at the end of that year, the venture had become jointly funded by private and public means and the cost had increased fivefold to A$1 billion.

In order to achieve this massive project, a number of legitimating and facilitating strategies were used. First, the project was taken out of the normal planning mechanisms of the Sydney City Council. The Darling Harbour Authority was brought into being by the Darling Harbour Act (1984) to be an authority with special development powers, which overrode normal planning requirements. Having adopted an effective mechanism facilitating construction and development, two simultaneous legitimation strategies were adopted. The first of these was the obvious one of a promised economic return from the investment both in the form of job creation and in the promised raising of revenue to be spent on health, welfare and public housing. In fact, the flow of funds went the other way, with a massive social redistribution graphically illustrated in the accompanying cartoon by Moir (see Figure 4.6).

The second legitimation strategy was a much more ideological and less tangible one. In the Premier's 1984 proposal, the Darling Harbour project was described as a 'Bicentennial gift to the nation' (Huxley and Kerkin, 1988). This concept of the Bicentennial gift, with its overtones of development for public benefit, allowed the project to be fast-tracked, minimised the consultation required and deflected attention from the social benefits which had originally been anticipated. It proved to have an acceptable and easily naturalised logic, following on from the development of Sydney Tower from 1978–81, successfully marketed both as Sydney's coming of age and 'the heart of the city' (Morris, 1982). Morris (1982:56) describes Sydney Tower

(continued)

(continued)

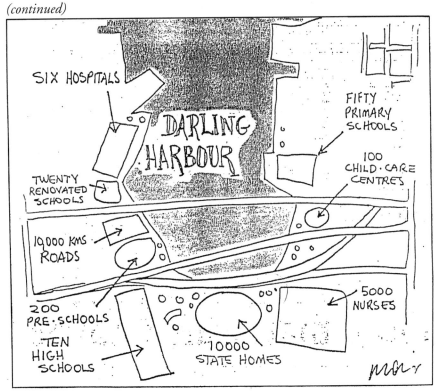

Figure 4.6 Cartoon of Darling Harbour
Source: Huxley and Kerkin (1988)

as 'an intentional, blatant and explicit production of a myth. It is a self-styled *artificial* myth' (her emphasis). As part of the continued redevelopment of the city centre, Sydneysiders were willing to take on board the mythological representations of both Darling Harbour and the Tower, as clear demonstrations that the city was visibly being incorporated into the global economy.

The 1988 Bicentenary of Australia provided the excuse to build a new and differentiated space of consumption. In the case of Darling Harbour, the consumer landscape was overlaid with elements of nostalgia, heritage and education. Similar major 'hallmark' events provide the legitimation for major construction by public authorities, whether these are anniversaries, world fairs or Olympic games (Ley and Olds, 1999). The construction of such landscapes forms part of the 'city of spectacle', often referred to as 'bread and circuses', in other words the food and entertainment which will keep the general public satisfied with their lot and not enquiring too closely into the social relations and financial deals behind such events. The development of Darling Harbour is just one of many such sites of consumption, which follow events and spectacles and which attract mobile and consumer capital in a continued quest for the differentiation of consumer landscapes.

the politically opposite Communards had been defeated underscored the symbolic triumph of the monarchists over the Communards.

The intersection of state and religion is apparent in yet other ways in the shaping of landscapes. In a number of religions, the world of the gods constitutes a macrocosmos on which the world of humans (the microcosmos) is patterned. This is usually the case when religion and its institutions exert significant power, so much so that normally secular practices such as urban planning and political action are subject to critical religious influences. Duncan (1990) illustrates this within the context of early nineteenth-century Kandy (modern-day Sri Lanka).

Political discourse in the Kandyan Kingdom was shaped by a number of texts outlining the reciprocal duties of a king and his subjects. These texts were drawn from an older set of Hindu and Buddhist political and religious texts such as the *Puranas* and the *Jatakas*. Two main narratives on kingship and landscapes prevailed. The Asokan narrative favoured the production of a landscape dominated by religious structures such as monasteries and public works such as irrigation tanks for the benefit of the people. The narrative was based on the Mauryan emperor Asoka (third century BC) who was considered an ideal Buddhist king. Such a king should be 'mild-mannered, righteous, and unfailingly protective of Buddhism and responsible for the welfare of his people' (Duncan, 1990:38). A king who successfully created such a landscape would gain political legitimacy. Contrariwise, the Sakran discourse follows after the Hinduised god-king whereby the king is seen as 'a kind of god on earth modeled upon Sakra' (Duncan, 1990:39), a *cakravarti*, a 'universal monarch who rules over his people and other kings just as the king of the gods, Sakra, rules over the thirty-two gods in the Tavatimsa heaven' (Duncan, 1990:40). Such a king favours a landscape of palaces and cities that glorify the god-king, modelled after the landscapes of gods on Mount Meru. A king's construction of such a landscape could gain him legitimacy as well, although Duncan illustrates how such 'rule' was less successful. In the early nineteenth century, the reliance on a Sakran discourse was confronted by contestatory readings of the landscapes and rituals by nobles and peasants, illustrating the 'volatile politics of interpretation' (Duncan, 1990:181). Both the landscapes of late nineteenth-century Sacré Coeur and early nineteenth-century Kandy thus illustrate the intersection of political and religion in shaping landscapes, gaining legitimacy for politics leaders and groups at times and causing a loss of power in others.

4.7 Summary

Landscapes are simultaneously medium and outcome of power. Because landscapes are everywhere in our everyday lives, they have central and significant roles in the playing out of power relationships. In both real and symbolic terms, they enable the exercise of power and, as we will illustrate in Chapter 5, the expression of resistance. Here, we have shown how landscapes make what is culturally and socially constructed seem natural, simultaneously naturalising the ideological systems that underlie their creation. We have illustrated how landscapes are inscribed with power in a variety of ways, for example, in the location and design of monumental structures, as in Canberra and Jakarta; in the segregation of social groups as in colonial cities, Vancouver's Chinatown and South Africa's townships and reserves; in the architecture and overpowering presence of particular building styles, as in New York's skyscrapers; in the controlled use of and meaning allocated to 'public' space, as in Hong Kong's commodified 'ambiguous' spaces, America's malls and Disneyland and various world's fairs; in the control of access to private space, as in gated communities; in the reinforcement of religious and political ideologies, as in Kandy and Sacré Coeur; and in the naming of and thus meaning allocation to streets and places, as in revolutionary France, post-colonial Singapore and former communist USSR. In all these illustrations across First and Third Worlds, capitalist and communist, contemporary and historical, we have illustrated that landscapes are shaped by the power of the state, of capital, racial and religious ideologies, and, often, the intersection of some combination of these.

4.8 Notes

1. The Commune of Paris was a revolt in Paris against the conservative provisional government established at Versailles following the defeat of the French in the Franco-Prussian War and the downfall of the Second Empire. Republican Parisians had sought to prevent a restoration of the monarchy and formed a revolutionary government. However, the Communards were defeated by government troops within two months.

Chapter 5

Landscapes of conflict and resistance

5.1 Conflicting ideologies, contested landscapes

In Chapter 4, we discussed how power relations may be expressed, maintained and enhanced through landscapes. At the same time, we intimated that contestations and resistances to power and ideologies can equally well be expressed through landscapes. In this chapter, we emphasise a reading of power relations which includes conflict and collision, negotiation and dialogue between groups. It has been argued by Foucault (1980a) in particular that power should neither be arrogated to the state or political system nor identified with particular individuals. Instead, Foucault has argued that power emerges from 'local arenas of action' and should be viewed as 'a "microprocess" of social life or pervasive feature of concrete, local transaction' (quoted in Agnew, 1987:23), 'run[ning] through the whole social body' rather than confined to the state or political system (Foucault, 1984:60–61). Its effects cannot simply be defined as repressive but may also be viewed as productive and enabling. Power is not conceived as a property but as expressed in specific strategies: its effects of domination are attributed to dispositions, manoeuvres, tactics, techniques, functionings. In short, this power is exercised and reproduced rather than possessed (Foucault, 1979).

We highlighted in Chapter 4 how political power gained and maintained through hegemony is more effective than that gained and maintained through force. However, Gramsci (1973) has also made clear that hegemony is never fully achieved, that is, those seeking to gain/maintain power will always be challenged in some way by other social groups. To attribute an absolute *omnipotence* to power would, as Colin Gordon (1980) has argued, confuse the domain of *discourse* with those of *practices* and *effects*. This is because what is intended and articulated by the 'powerful' within the domain of discourse (such as that of a nationalist ideology) may fail to materialise in its entirety when transposed to the domain of actual practices or they may produce unintended consequences and effects. Further, to attribute omnipotence would also amount to denying that those which disciplinary power seeks to control are capable of counterstrategies which can challenge disciplinary power and modify its effects. As Giddens has contended, 'no matter how great the scope and intensity of control superordinates possess, since their power presumes the active compliance of others, those others can bring to bear strategies of their own, and apply specific types of sanctions'. Foucault (1980b:142) too argues that there are possibilities for 'revolts to the gaze' and contends that resistances to power 'are all the more real and effective because they are formed right at the point where relations of power are exercised [and that] like power, such resistance is multiple and can be integrated in global strategies'.

Specific forms of resistance may be overt and material but they could as well be latent and symbolic. In other words, while resistance represents political action, it can be conveyed in cultural terms, for example, through the appropriation and transformation of the material culture of the dominant group (see, for example, Hall and Jefferson, 1976; Hall *et al.*, 1978). This is often the means of resistance adopted by the 'weak' (Scott, 1985), and takes the form of 'tactics' (de Certeau, 1984) and 'rituals' (Hall and Jefferson, 1976) rather than 'strategies' (de Certeau, 1984; Hall and Jefferson, 1976). 'Strategies of resistance' suggest some degree of reflection over past circumstances and anticipation of future ones, allowing a 'rational' decision to be made; some choice between alternative courses of action; some exercise of power in choosing between options or a response to a lack of power. 'Rituals of resistance', by way of contrast, are shorn of the deliberation associated with strategies and suggest routine and somewhat unreflective acts. Some of the recent literature about the cultural turn in geography suggests that the focus on cultural rituals, tactics and the like takes the attention away from 'real politics', positioning cultural politics as a 'weaker' or less effective politics (see Chapter 2). This has been characterised by Badcock (1996) as a redirection of attention from 'messy political practice' and the 'real problems of cities' to the 'politics of representation'.

Hegemonic control and resistance are constituted both at the discursive and practical levels of social reality. Conflict and negotiation may occur as a result of the collision of discourses, but more often than not, exist at the level of habitual practice as part of everyday life. It is in daily encounters that power can be studied 'in its external visage, at the point where it is in direct and immediate relationship with . . . its object, its target, its field of application, . . . where it installs itself and produces its real effects' (Foucault, 1980c:97). It is at this level too that subjugation is achieved, challenged or inflected. The 'concrete space of everyday life', to borrow Henri Lefebvre's term, is not only enframed, constrained and colonised by the disciplinary technologies of power, but also 'the site of resistance and active struggle'. The commonsensical, pragmatic circumstances of everyday life not only provide the context for the 'experience of culture' (Cohen, 1982) but in as much as everyday encounters are embedded within and shaped by the larger discursive/political context, ordinary people in turn affect political discourse, certainly through their inscriptions on landscapes. Indeed, the centrality of landscapes in such encounters cannot be overemphasised, as we pointed out in Chapters 3 and 4. In as much as the powerful leave their imprints on landscapes, the oppressed can and do challenge the powerful ideologically through built forms which they see as unjust and buildings have been symbolically destroyed and/or occupied in various struggles: student protests, opposition to despotic government regimes and race relations, for example (Laws, 1994:9). But the oppressed are not the only ones who challenge and resist, for the powerful elite who attempt to maintain the status quo or to keep the non-elite 'out' will also oppose and contest landscape developments that threaten their worldview and landscape experience. In this chapter, we examine the role of landscapes in the expression of conflict and resistance. We use a range of examples, many of which are to do with our everyday landscapes. Streets – including street names and street life – are the substance of many illustrations. The reclamation of streets by carnival, protests and parades help us to make the point about landscapes as sites of conflict and resistance. Neighbourhoods and the communities' resistances against 'pollution' through the presence of prostitutes, homeless, bikers, minority groups and toxic industries also place the accent on how everyday landscapes are sometimes riven by conflicts. Beyond these quotidian examples, we also draw attention to larger scale conflicts, such as those anchored in religious strife (Ireland) and linguistic and cultural divergences (Sri Lanka).

5.2 Symbolic resistance

Resistance occurs in a variety of symbolic ways for different ideological reasons. First, symbolic resistance may be understood in terms of landscapes of the spectacular and extraordinary. Spectacle and extraordinariness need not represent resistance in and of itself. Yet large-scale temporary movements of people, transforming ordinarily sedate landscapes, altering the normal order of things by the sheer numbers and the activities they engage in, could signal resistance in a symbolic way to ordinariness and 'normality'. Carnivals, parades and marches are clear examples of such symbolic resistances and are effective precisely because they are so out of the ordinary. We will elaborate with examples later. That said, however, symbolic resistance need not involve spectacle and the extraordinary. Everyday, familiar and prevailing landscapes may embody resistance or represent conflicting ideologies and affiliations. The examples of street names and cemeteries illustrate the point, where the act of naming and the language used become symbolic of deep-seated resistances. One way or the other, the different symbolic modes of resistance are prompted by different ideological reasons: divergent racial and gender ideologies and practices, differential access to economic and political power, distinct political affiliations and contrary national allegiances and senses of community.

One example of symbolic resistance is carnival and the concomitant temporary inversion of normal hierarchical relations. Following Stam (1988:135), Folch-Serra (1990:265) lists the characteristics of carnival that constituted it as a radical activity. Carnival is construed as the 'space of the sacred' and as 'time in parenthesis'. Its main features are ambivalence, festive laughter and parody. The ordinary order is suspended, as is everything else resulting from socio-hierarchical or any other form of inequality. In turn, there is free and familiar contact among people who are normally separated by impenetrable hierarchical barriers. Taken to an extreme, there is in fact social inversion, a counter-hegemonic subversion of established power. As Folch-Serra (1990:265) summarises:

> In laughter as a spectacular feast of inversion and a parody of high culture, Bakhtin sees the possibility of a 'complete withdrawal from the present order', of a carnivalesque game of inverting official values where antihierarchism, relativity of values, questioning of authority, openness, joyous anarchy, and the ridiculing of all dogmas hold sway.

In these 'carnivalistic mésalliances' where the lofty is combined with the low, the great with the insignificant, the wise with the stupid, carnival also serves as 'a pageant without footlights and without a division into performers and spectators' (Folch-Serra, 1990:265). In distorting the relationship between performers and spectators, carnival is no longer a spectacle seen by people; indeed, everyone participates in it and there is no life outside it (Jackson, 1988:225). Carnival is a provisional landscape for a 'second life' (Bakhtin, 1968:9), an escape from and a critique of the static, oppressive hierarchy of normal class and economic relations. It is a space for change and disorder and 'temporary liberation from the prevailing truth' (Bakhtin, 1968:10). It embodies temporary rebellion not only of the lower classes but of the lower faculties: of instinct versus reason and flesh versus spirit. Indeed, carnival celebrates fluidity and change, mocking the forms which embody an illusion of stable hierarchy. The landscape of carnival is a landscape of change, inversion and even rebellion.

By way of contrast, Jackson (1988:216) points out another logic to the understanding of carnival. The symbolic reversal of social status should not be confused with subversion because it only serves to reaffirm the permanence of the social hierarchy outside the time and landscape of the carnival. What carnival offers is a temporary respite from normal relations. In this way, carnival is a 'social leveller', allowing for 'a harmless release of tension, and a force for social integration' (Jackson, 1988:215). This is a somewhat 'dismissive' interpretation of carnival, suggesting that the politics of carnival are superficial. Bearing in mind both understandings of the carnival, it is obvious that carnival landscapes can have at least two functions. They at once represent symbolic resistance and, concomitantly, a temporary inversion of normal hierarchical relations.

In a study of West Indian identity in Britain, conducted through an analysis of the annual carnival in Notting Hill, Jackson (1988) illustrates how carnival is a contested social event whose political significance is inscribed in the landscape. The event reflects the spatial constitution of symbolic resistance, achieved through the symbolic reversal of social status. This offers a temporary respite from normal relations of subordination and domination. Indeed, it plays an integrative function in creating a specifically 'West Indian' identity in Britain and, thus, a potential platform for protest, opposition and resistance (Jackson, 1988:222). It is perhaps for this reason that the authorities cast carnival as an event that must be kept to the road because of its potential as a threat to the social order (Jackson, 1988:223). This also reflects the territorial basis of British racism. The implicit resistance represented by carnival is made explicit in the riots of 1976, which offered 'an opportunity for the ideological construction of "black youth" as an implicitly male, homogeneous, and hostile group, leading to the subsequent "criminalization" of black people in general' (Jackson, 1988:214). To that extent then, the anti-hierarchism that signals symbolic resistance is indeed temporary and should not be confused with subversion because it only serves to reaffirm stereotypes and solidify social hierarchies (Jackson, 1988:216).

In as much as carnival as symbolic resistance casts light on race and class relations, so too it does on gender relations. Lewis and Pile (1996) argue that the Rio de Janeiro Carnival opens up possibilities where the performing and masquerading of femininity become a site of resistance. This is achieved through the introduction of 'uncostumed' female paraders, uncostumed in that they are not naked, but display many parts of the body while dressing certain parts. These female bodies become available for lascivious, desired, sex-based scrutiny by men. They break the rule of private, domestic femininity, thus transgressing the rules of chastity. The essentialised notions of this private, domestic femininity are confused and challenged when the female body is in effect performed at the carnival – used by the women and sometimes men (transsexuals or transvestites) to captivate and manipulate the dominant gaze in order to affirm the woman's desirability. Such manipulation of female desirability is most overt in the case of female stars of past Rio carnivals who then take the leap from poverty to success, leaving behind their shanty town backgrounds, starting their own business in fashion, cosmetics and modelling and starring in films. They have thus taken advantage of the contradictory social values to achieve success in the hierarchical world of business.

In contrast, Spooner (1996) examines black women's experiences of the annual African–Caribbean carnival in St Paul's, Bristol, in the UK, and highlights how carnival does not necessarily enable the contestation of regular gender roles in the African–Caribbean community. While it is tempting to conclude from various evidence (the fact that women dominate many stalls selling food at the carnival and are centrally involved in the organisation of the carnival) that they have entered the male-dominated productive sphere, thus inverting or subverting everyday gender roles, that is not the case. This is because those actions are extensions of the everyday in these African–Caribbean communities, where women are often engaged in petty businesses (unlike the white, middle-class experience). This highlights the danger of assuming that experiences of the white, middle-class can be used to help understand those in other communities.

Apart from carnival, other kinds of parade can also have political significance, reflecting conflicts between social groups with differential access to political and economic power. While our preceding discussions of carnival focus on each landscape as the site of conflict and subversion, in some situations, resistance is achieved by the introduction of an alternative event, in this case, an alternative parade 'conceived and staged in conscious opposition to a community's dominant or primary public ceremony' (Lawrence, 1982:155). The Rose Parade in Pasadena, California, is a case in point, occurring since 1890 on an annual basis, but faced with the ritualised protest of the Doo Dah Parade since 1977. The former is designed to be a focus of civic pride, constituting one of the key elements of Pasadena's urban identity. While comprising bands, beauty queens, equestrian units and other such elements as may be found in many American parades, the Rose Parade is distinguished by huge florally decorated floats, submitted by corporations, voluntary associations and politico-administrative units. Participants are carefully selected to represent local communities, regions of the country and other countries. Controversial themes are avoided and some celebratory aspect of American culture is usually chosen, such as 'The Great Outdoors' and 'Heritage of America'. These themes approximate ' "middle class American standards of good taste": patriotism but not politics, God's presence but no particular religion, feminine beauty but not sex' (Lawrence, 1982:161). These messages are meant to be clearly innocuous and leave little room for alternative interpretations. In terms of process, all float designs must be approved by the organisers. The organisers themselves come from the Tournament of Roses Association, a voluntary organisation whose membership is exclusive (one needs to apply to be a member and be subject to a selection process) and hierarchical (members must serve for several years in various capacities of ranked importance and prestige before being appointed to positions of authority and responsibility). The Rose Parade, Lawrence argues, is defended by local residents with almost sacred reverence and is regarded as the focal point of its identity as represented to the world beyond Pasadena. It seeks to unify by avoiding divisive issues and through the collective enforcement of rules regarding theme/content and clarity of message.

The Doo Dah Parade is an explicit alternative to the Rose Parade. It emphasises spontaneity as opposed to the rigidity and conservativeness of theme and implementation of the Rose Parade. The organisers (originally patrons of a bar, mainly artists who lived and worked in Pasadena's 'Old Town', making up Pasadena's counterculture) intentionally refuse to set up rules and regulations: hence, 'no theme, no contests, no prizes, and no parade order' (Lawrence, 1982:164). The intent was to admit anyone who wished to to participate and to encourage the most 'zany, eccentric, and outrageously creative ideas' (Lawrence, 1982:164) and for participants to interact freely with the audience. Doo Dah also relied on the informal networks of friends rather than the formal hierarchy of the Tournament of Roses Association. In part, the Doo Dah organisers and participants were guided by a concern about the encroachment of urban redevelopment in the Old Town and the invasion of high-rise office buildings which threatened eviction of the older community. The location of the Doo Dah Parade in the Old Town was not incidental, but designed to showcase the vitality of the community and arouse interest in its preservation. At the same time, the thematic content of entries expressed political protest against a range of issues (e.g. 'Immoral Minority') and consciously also mocked elements of the Rose Parade (e.g. 'The Torment of the Roses'). However, these messages are often so symbolically embedded and expressed in such an ambiguous way that it is possible for some to enjoy a performance with the full meaning slipping by, while for others the satirical intent can be provocative. The Doo Dah thus represents a ritual of rebellion against the established order, in particular, resistance against an elite organisation controlling the symbolic image of Pasadena, its economic fate and development plans.

The Mardi Gras parade in Sydney is an excellent example of resistance turned mainstream. When it started in the late 1970s, homosexuality was still illegal and the first parade in June 1978 consisted of less than a dozen floats, literally battling police. As the parade developed after homosexuality was legalised in 1984, so the floats became more numerous and more daring (Seebohm, 1994). The organisers developed the location of the parade to ensure its symbolic significance, starting outside the gay area of the city under the twin sacred and secular legitimating authorities of the Cathedral and the Town Hall, passing through the gay hub of Oxford Street and ending again in mainstream Sydney. The pathway emphasises the (temporary) normality of the gay in a carnivalesque inversion of heterosexual social relations. The parade has been made more inclusive by its designation as the Gay and Lesbian Mardi Gras since 1995. By the late 1990s, the Mardi Gras had become the biggest annual tourist event/attraction of the state of New South Wales, bringing in some A$14 million annually. As the Mardi Gras has become more inclusive and more accepted, so some members of the Sydney gay community are attempting to maintain its exclusivity (of at least some of the events) to gays and lesbians (Seebohm, 1995). What started as a ritual of resistance against gays being excluded has not only become mainstream but has also become a means whereby they maintain the exclusiveness of their gay identity. Nonetheless, the parade itself is entertaining for hundreds of thousands of participants and for millions more through television and video coverage.

Other means of public symbolic resistance are less entertaining than carnivals and parades and more openly and explicitly confrontational. Take Back the Night or Reclaim the Night marches are one example, representing symbolic statements against crimes targeting women. The slogan 'Take Back the Night' was first used in the United States during a protest march down San Francisco's pornography strip in 1978. It took place at night, symbolically to protest against perpetrators of violence and crimes against women, which though not necessarily time specific, also included many which took place particularly under cover of night. Such crimes – e.g. rapes, muggings and sexual harassment – were recognised to be linked to portrayals of women in different media and arenas, from sexist articles to advertisements and pornography. The combined effect is that women are unable to walk the streets at night without a male for protection. The first Take Back the Night march sought to address this inequality and involved over 5000 women from 30 American states who then returned to their own communities to continue the work started in San Francisco, calling for perpetrators of violence against women (including rapists, batterers and pornographers) to be held accountable for their actions (Lederer, 1980).

Just as street names can be an expression of power (Chapter 4), they can also be symbolic expressions of conflict and resistance. Azaryahu (1996) cites various examples. In 1988, when a prominent Palestinian Liberation Organisation leader, Abu Jihad, was killed, his name was used in improvised street signs in the West Bank, as a way of commemorating a national martyr and defying Israeli rule. The Israeli authorities quickly removed the unauthorised street signs. In the case of Lodz, a Polish city which could only resist communist rule symbolically, local residents defiantly refused to use the 'communist' street name, Street of the Siege of Stalingrad, abiding by the former name, Street of November 11th, evoking the memory of the independent Polish republic of the pre-war period. Indeed, as Lefebvre (1991:54) argues, the creation of new spaces out of conflict, in particular, revolutions, is a necessary condition of successful resistance:

> A revolution that does not produce a new space has not realized its full potential; indeed, it has failed in that it has not changed life itself, but has merely changed ideological superstructures, institutions and political apparatuses. A social transformation, to be truly revolutionary in character, must manifest a creative capacity in its effects on daily life, on language, and on space – though its impact need not occur at the same rate or with equal force, in each of these areas.

As Azaryahu (1996:318) points out, renaming streets and other public spaces has an immediate impact on daily life, language and space and illustrates a 'radical restructuring of power relations', particularly when it occurs after revolutions. An example is the massive renaming of streets in West and East Berlin that coincided with changing political circumstances: movement from a monarchy to a republic, the rise of Hitler to power, the surrender of Nazi Germany, national division and reunification. Between May 1945 and December 1951, for instance, 200 streets in West Berlin and 227 streets in East Berlin were renamed (Karwelat, cited in Azaryahu, 1996:318). Between April and May of 1951, 159 East Berlin streets were renamed to decommemorate the Prussian tradition (Azaryahu, 1992). Similarly, Pred (1992) illustrates how dominant ideology in the 1890s prompted official street name revision to reflect patriotic and historical names, Nordic mythology, famous Swedish authors and prominent men in technology and engineering. Yet, the working class continued to employ previously used names, underscoring working-class resistance to middle- and upper-class ideology in nineteenth-century Stockholm.

5.3 Overt conflicts, contested terrains

Conflict between ideological positions may move beyond the realm of the symbolic to overt conflict, involving a range of expressions, differing in degrees of confrontation and violence. Such conflict is often expressed through struggle over control of landscapes. Through such conflict and resistance, different sites of everyday practice may be understood as assertions of identity and struggles against divergent values. While power relationships are often more complex than can be neatly classified, for current purposes, three broad categories of relationship will be examined: first, the ways in which the more powerful oppose the incursions of the marginal, through which landscape is a way of marginalisation and exclusion; second, the ways in which the less powerful seek to assert their preferences, resisting the impositions of the powerful through attempts to create or maintain their own landscapes; and, third, a continued balance of power which gives rise to intractable conflicts over territory.

5.3.1 Resisting and excluding the marginal

Instances of the powerful resisting the incursions of the marginal are best ex-emplified in some forms of neighbourhood activism, with the more powerful stable neighbourhood residents seeking to preserve self-identity and exclude 'threatening others' by opposing the presence of various groups, such as the homeless, prostitutes and prisoners, the mentally ill, the disabled and people with AIDS. Overt conflict has often crystallised around resident action groups, communities of people who seek to keep away certain facilities and services from their neighbourhood. They almost invariably acknowledge that such facilities are needed, but approve of their construction elsewhere, a syndrome termed NIMBY (not in my backyard). The desire to keep the 'Other' away is as much about preventing negative externalities (such as the negative effect on local property prices) as they are about a desire to maintain cleanliness and purity (Sibley, 1995). Sibley argues that this

> fear of the self being defiled or polluted is projected (or mapped) onto specific individuals or groups who are depicted as deviant or dangerous. . . . [S]patial exclu-sion has been the dominant process used to create social boundaries in Western society, and the key means by which hegemonic groups, normally white, middle-class and heterosexual, have been able to marginalise and control those who do not match their ideas of what is an acceptable way of living or behaving (Hubbard, 1998:281).

Public space is therefore a space of exclusion, rather than the idealised democratic space in which people can exercise their rights of citizenship (Mitchell, 1996). We discuss the examples of resistances against the homeless, homeless service providers and prostitutes to illustrate these arguments. In the case of the homeless, the homed feel threatened by the presence of the home-less and the 'blight' that their presence indicates. Their general success is then achieved by marginalising the homeless using laws, regulations and various forms of policing and cleansing. We then illustrate how residents oppose the presence of prostitutes in their neighbourhood because of the perception that the moral order is disrupted. Their fears for their safety in this case are centred on fears of moral corruption. Fears for physical safety are also inflamed by the threat of the 'demented' homeless in their midst.

The conflict and resistance exemplified by the homed ('people on the inside') and various homeless ('people on the outside') in Los Angeles reflect the ways in which the powerful (the homed, with economic resources and access to planning guidelines and legal reconstructions) seek to control public space and community resources, representing social polarisation of the 1980s (Wolch, 1995:78). The homeless, cast as the 'Other', are seen as impossible to integrate into society because of their differences in origin, appearance and behaviour. Indeed, not only do the homed construct the homeless as 'different', academic research has at times (inadvertently) cast them as urban nomads and postmodern primitives who, like hunters and gatherers forage to secure their needs (Wolch, 1995:81). Worse, epidemiological research has tended to medicalise and psychiatrise homelessness, creating the image of the homeless as 'different, diseased, or demented', thus contributing to feelings of threat where they exist. As a result of these various ways in which the homeless have been constructed and understood, innumerable cities and neighbourhoods have obstreperously defended their local turf against incursions by the homeless, in an attempt to 'cleanse' the landscape. For example, in Skid Row in Los Angeles, homeless sidewalk encampments are routinely and forcibly removed. Wolch (1995:81) describes it thus:

> Street sweeps routinely involve police to roust sidewalk dwellers, a skip-loader to remove belongings and structures on the sidewalk, and the street sweeper itself. Equally subtle are the sprinklers attached to the walls of many businesses, which randomly shower the sidewalk with water to drive homeless people away. And in the suburbs, the exclusionary milieu is perfectly captured by the proliferation of privatized police and defended spaces.

In 1987, for example, Los Angeles Police Department (LAPD) began a systematic effort to clear Skid Row streets of 'cardboard condos'. The Central City East Association (CCEA), an association comprising the wholesale toy and fish-processing industries next to the homeless and social services agencies on Skid Row, also lobbied hard for a greater LAPD presence in the neighbourhood (Goetz, 1992:548). Further, the homed also sought to contest use of public space by the homeless by containing these 'offending people' and their activities within clearly defined areas. Planning regulations were rewritten to make illegal camping or sleeping in public places, such as parks, under freeway overpasses or near railway tracks. In Miami, for example, this strategy of containment, in which homeless shelters and services are relegated to particular circumscribed places at extremely high densities, has been said to approximate apartheid. As Wolch (1995:81) argues, 'homeless people are offered "safe zones" (or homelands?) but then forego rights to exist in any other part of the city's public realm'.

The homeless, however, do not constitute a homogeneous group, as Sue Ruddick (1996) illustrates. In a study of homeless young people in Hollywood from 1975 to 1992, she shows how runaway and homeless young people did not gravitate to existing concentrations of homeless people, for example, in Skid Row, but had moved instead to Hollywood. They reproduced punk subcultural practices and occupied cemeteries, clubs and empty manor houses. Their use of such space was also contested by the community, which expressed hostility towards their presence as well as the services that had begun to locate in and around the Hollywood area to cater to the target group. Such resistance to their presence was ameliorated by service providers who sought to draw attention to the fact that the homeless young people were victims of family breakdown, forced into occasional criminal activity in order to survive, rather than 'criminals'. However, most commonly, the experience is that homeless services providers have had to engage in bitter disputes with local community residents, resulting in the frustration of efforts to establish facilities for the homeless. The politics of turf thus cause the homeless to continue in a 'cycle of exclusion . . . from legitimate or adequate space within the city' (Gleeson and Wolch, 1989, quoted in Costello and Dunn, 1994:63).

The resistance offered by homeless girls in the New South Wales city of Newcastle shows exclusion by the authorities, ritualised resistance to the police and codified behavioural resistance to gendered norms. A small population of homeless young people took up residence in accommodation either vacant as a result of industrial change, such as the derelict wharves along the waterfront or in housing damaged by the unexpected earthquake of 1989 (Winchester and Costello, 1995). Their presence was certainly unwelcome to the city authorities who were pleased that the proposed redevelopment of the city waterfront would force both the relocation of the homeless themselves and of their service providers. In more than one case, the homeless children resisted authority in a ritualised game-like form, setting spot fires and enjoying the spectacle of the police racing around the inner city to deal with them (Winchester and Costello, 1995). Furthermore, they formed their own subcultural identity with a moral order and codes of behaviour which emphasised group solidarity and protection, while resisting gendered domestic behaviour and norms of bodily attractiveness. They would group together to defend one of their own from attack or set out to look for and assist two new girls who had recently appeared on the scene and in so doing formed a bond within the group more effective than the families that had failed them (Winchester and Costello, 1995).

Prostitution and prostitutes are popularly considered to be producers of negative externalities for the homed and comfortable. Here we examine the geographies of street prostitution in two British 'red light' districts (see also our discussion of landscapes of prostitution in France and the Philippines in Chapter 6). Hubbard (1998) illustrated how campaigners in Manningham in Bradford and Balsall Heath in Birmingham, opposed the presence of female street prostitutes from those areas through protracted pickets. In the case of Manningham, Bradford, community activism was ignited because residents (primarily south Asian males) felt that existing regulation of street prostitution did not protect them from street prostitutes and kerb crawlers. Provocation came in the form of prostitutes and their clients using the car park of the neighbourhood mosque. Direct neighbourhood protests took the form of community watch-style pickets, placards and street patrols, led by elders of the mosque, designed to disrupt the work of street prostitutes. Registration numbers of kerb-crawlers' cars were taken and handed to local police. Although existing neighbourhood groups had previously called for more punitive policing, this campaign involved much more direct action and did, on occasion, lead to harassment and confrontation, even though the original intent was peaceful protest. Pickets claimed to receive threats from pimps and associates of prostitutes. At the same time, prostitutes alleged verbal and physical abuse. Nevertheless, the campaign began to yield results, with a dramatic drop in prostitution, to the extent that the local police co-opted the campaign, asking that all pickets register with them to form a 'neighbourhood watch'-type group.

The confrontational and exclusionary tactics and strategies to keep the moral order of the streets are anchored in gendered and racialised/religious turf politics. They are rooted in the social construction of prostitutes as deviant 'Others', 'dirty, devious, dangerous and diseased' (Hubbard, 1998:282):

> offering sexual services outside the socially-sanctioned institutions of domesticised monogamy and ostensibly disrupting the moral order of society. Hence, although the nuisance and noise created by prostitutes and kerb-crawlers was stressed by a number of residents as the cause of their protest . . . , it is difficult to accept that the level of this nuisance alone would be sufficient to motivate the widespread community protests witnessed (Hubbard, 1998:281).

The case illustrates well the social constructedness of a category – 'prostitutes' – and how landscapes and, in this case, streetscapes, shore up these ideological constructions of morality.

A third example will throw into profile the ways in which resistance based on religious prejudices is designed to exclude the marginal 'Other'. Muslims in Sydney have found it very difficult to build places of worship. Councils and resident activist groups have frustrated the public practice of Islam and other non-Christian faiths. Many have been forced to worship secretly, in residential properties or to use commercial premises. In 1996, there were 96,792 Muslims in Sydney, whereas in 1976, there were only 22,206 (Dunn, 2001:293). The pressure for mosque building was therefore particularly acute within Sydney in the 1980s and 1990s.

The significant presence of Muslims in Sydney notwithstanding, opponents of mosques have consistently asserted that there were no local Muslims. Resident objectors argued that members of Islamic groups proposing the developments were 'outsiders'. Muslims were constructed as the non-locals, as 'they' or 'them' (Dunn, 2001). Opponents of mosques often made very direct claims to local citizenship. This was achieved by describing and identifying themselves as 'citizens'. Letter writers would sign their objections or begin sentences with descriptions such as 'concerned citizen', 'concerned Christian', 'legitimate resident', 'local' and 'rate payer'.

Mosque opposition and anti-multiculturalist discourses were intertwined in Sydney. Opponents asserted that mosques should be refused consent by local authorities and Australian Muslims should give up Islam and become 'Australians', by which they meant Christian (Dunn, 2001:303–5). Multicultural and multi-religious existences could only be construed as detrimental to Australian culture, as one letter writer put it:

We are tired of our government selling Australia out to people such as these and then letting them impose their way of life on us instead of them learning to be Australians (protest letter to Campbelltown City Council, 8 August 1990).

These quotations utilise a series of Self/Other binaries, in which 'our', 'Australians' and 'non-Muslim' are the Self, while 'they', 'non-Australian' and 'Muslims' are the Other. Opponents of mosques and Islamic centres drew on a nationalist ideology of 'Australia as Christian'. And they complained about the threat to 'their' hegemony:

The families living in the Green Valley/Hinchinbrook region have been accustomed to enjoying a peaceful and harmonious life with their fellow moral and God fearing Christians (protest letter to Liverpool City Council, 10 August 1989:3).

According to these constructions of the Australian nation, Islam did not belong. There was no place for non-Christians among these Christian landscapes.

These examples illustrate the politicisation of landscapes and their meanings, through the use of landscapes by the powerful to keep out the marginal and the 'Other'. While social constructions of the 'Other' are apparent in the discursive realm, in debates, in pronouncements and in texts, landscapes concretise and manifest the power of the dominant to resist the incursion of the 'Other'.

5.3.2 Resisting impositions of the powerful

As we illustrated in Chapter 4, the impositions of the powerful are everywhere evident. Frequently, the less powerful seek to contest these impositions and resist changes wrought on their landscape. Such contestations may involve direct action and outright defiance; legal action; or attempts to work 'from within'. At other times, resistance is diffused and poorly coordinated, resulting invariably in failure to fend off the powerful.

Overt confrontation and open defiance of powerful ideologies is evident in the ways in which local populations in some Third World countries have responded to impositions flowing from ideologies of what constitutes nature. Whatmore and Boucher (1993) offer a useful overview of the different meanings (ideologies) that underlie conflicting attitudes towards nature. These are: a scientific narrative which invests nature with value as a scientific repository, realised in a system of nature reserves; an ecology narrative which draws from a scientific base and takes a position that land development is a potential threat to the environment, thus reinforcing a regulating planning ethos; a narrative of aesthetics which interprets nature as a source of aesthetic value, realised in the form of national parks; and a commodity narrative which interprets the land development process as one in which new and better environments are produced – cast as products – such as nature reserves, country parks and, in the context of leisure, residential and commercial developments, landscapes and planted trees (see also Box 2.5 on the multiple meanings of landscapes).

In this last narrative, the distinctions between the 'natural' and 'built'/'created' environments are blurred.

The various ideologies have resulted in conflicts over land use negotiated in different ways towards resolution. The conservation movement and its associated values has attracted contested narratives, for example. Wood (1995) argues that conservation can sometimes be insensitive to local needs. In such a situation, the majority of local people view wildlife conservation as 'alien, hypocritical, and as favouring foreigners'. He cites the example of conservation efforts in African environments, which he argues is an alien transplantation from the North American construct of national parks, a 'rich country institution' (Wood, 1995:118). Yet, it is no more than an act of robbery – taking from indigenous peoples their homeland and 'assigning it to an artificial idealised landscape in which humans have no place' (Wood, 1995:123). Worse, the traditional uses of resources by local communities, such as hunting, woodcutting and grazing, have been labelled as 'illegal'.

In more recent years, conservation movements have been paralleled by the exoticisation, packaging and promotion of Third World nature attractions by tour companies, particularly those in the west. The initiation of most wildlife conservation and tourism projects in Kenya, for example, is spurred on by western conservation and development organisations (Akama, 1996). Wildlife conservation and tourism development policies and programmes in Kenya are, in turn, influenced by the values and desires of the western world, with the creation of national parks and the development of tourism hotels and lodges, 'justified' by the fact that western tourists make up the main clientele. These 'pristine environments' represent alternatives for escape from the stresses associated with (particularly western) urban life and industrial capitalism (Akama, 1996:569). While the development of national parks and their aestheticisation has been prompted primarily in the west by large-scale urbanisation and industrialisation which has provoked a search for nature, the 'Other', the population in Kenya and many other Third World countries are primarily rural. These rural communities living around protected natural areas have different social and environmental values, concerned mainly with meeting subsistence needs amid poverty, famine and wildlife destruction of their property. Even the small urban middle class in Kenya has different roots from that in the west. Most have familial and social ties to the rural areas, which they see to be places of ancestral origins rather than as exotic environments for wildlife observation.

Resistance to these powerful ideological and material actions represent the behaviour of those who have no other recourse but open opposition. From the position of distinct powerlessness, rural communities, who have little or no influence over policy making, express their divergent views and resistance by their continued presence in national parks. As one elderly man in Kenya's largest national park expressed: 'Whether they [the government] like it or not, we are going to occupy this wilderness. This land used to belong to our forefathers' (Akama, 1996:572–3). Landscapes in this instance are more about the source of identity and belonging and less about conservation and tourism, illustrating once again how they are very much socially constructed.

This refusal to be re/dislocated as a strategy of resistance is adopted in urban areas as well. The contestation over space between the minority Sikh community in Singapore and the secular state which emphasises developmentalism provides such an urban parallel to the conflicts over natural landscapes in Kenya. Kong (1993) examined the circumstances surrounding the Sikh community's resistance against change imposed by the state and the manner in which the conflict was 'resolved'. In 1978 the Sikh community's central temple site was to be acquired because the area in which it stood was largely occupied by pre-war shophouses and was due for urban renewal. A Ministry of National Development statement pointed out that comprehensive development compatible with good planning is not feasible if the site occupied by the temple were to be excluded. Hence, the land was to be compulsorily acquired to make way for four blocks of flats. Although there was no compulsion for the government or its agency (the Housing and Development Board-HDB) to provide alternative sites for religious buildings affected by public projects, it made an exception in this case because the temple was the main Sikh temple catering to all sects. The government therefore offered the community other sites for its relocation. A first site was offered in June 1977 but was turned down. A second site was subsequently offered in February 1978, which was again rejected. At that point, community leaders called for a meeting attended by 750 members, at the end of which three resolutions were passed. The first was that the state should preserve the temple as evidence that it was treating all religious groups even handedly. Second, they urged the government to protect the religious rights and interests of the Sikh community. Third, they took a unanimous resolution that the temple should remain on the Queen Street site; there was no question of accepting an alternative site or selling or exchanging it. The resolutions were sent to then President, Henry B. Sheares, and then Prime Minister, Lee Kuan Yew. In May 1978 the Prime Minister met with nine Sikh community leaders and stressed two points. One was that all religions were and would continue to be treated equally. The other was that Singapore's progress through redevelopment had to go on, meaning that the old had to give way to the new and this included churches, temples, mosques and so on. He also managed to obtain agreement from the Sikh representatives that there had been no discrimination against the Sikh community in Singapore. After the meeting, a spokesman for the group expressed confidence that the matter would be resolved amicably soon, as the Sikh community also wanted to contribute to the progress of the nation. Eventually, the community moved out of its premises into temporary buildings until 1986 when members moved into a new temple in another site.

Whereas open defiance did not work in the Singapore case, negotiation gave way to direct action in the case of Dayton, Texas, eventually resulting in the success of the community concerned in achieving desired outcomes. Dayton is a small, working-class community, comprising about 6000 people. It was identified as a strategic site for a waste disposal company, Hunter Industrial Facilities, Inc. (HIFI), because of its proximity to Houston, with its petrochemical industry. The proposal was to construct ten storage caverns for hazardous wastes, each equivalent in height to the Empire State Building. Residents were naturally concerned and opposition to the site emerged. Recognising this, HIFI invited them to form a committee of local leaders (the LRC: Local Review Committee) to discuss the community's apprehension and to negotiate conditions for the site. To that extent, HIFI was strategic in its move. Structured discussion formats and negotiations have been shown to advantage corporate representatives at the expense of residential communities, with dangers ranging from intimidation and undue influence to manipulation, often because corporate resources enjoy greater experience, technical knowledge and finances. Residents may 'become hostage to the corporation's power and experience the "Stockholm syndrome", in which hostages take on the ideology of their captors' (Murphree et al., 1996:448). However, in this particular case, cooption did not work.

Such cooption was attempted in various ways. First, conflict was institutionalised in the sense that formal associations are formed and recognised as the proper and orderly channel through which views may be expressed in both directions. In this way, opposing parties can be more easily and efficiently handled and neutralised. Inviting the formation of LRC was HIFI's first step in this direction. To achieve some degree of legitimacy and credibility, key leaders in the opposition rank must be included in any such formal association. The LRC included, inter alia, the local environmental group (PACE: People Against a Contaminated Environment). Second, such cooption ensures that groups resisting any proposal feel included in the decision-making process, even though their recommendations may not actually be taken on board. Third, cooption is successful when residents' concerns appear to be adequately addressed, a process termed salience control (Murphree et al., 1996:457). This is where the process broke down in Dayton. Residents began to feel that the LRC was not representing them sufficiently and, indeed, appeared to be defending the proposal on behalf of HIFI to those opposing. Members of the LRC appeared too friendly and familiar with HIFI officials, which residents interpreted to indicate endorsement of the proposal. Accusations of cooption surfaced, as PACE activists were variously alleged to have accepted money from HIFI in the face of personal financial hardship or to have accepted work from HIFI. The ultimate evidence of collapse in faith in the LRC occurred when another opposition group, FACE (Families Against a Contaminated Environment), was set up, with a massive defection from PACE. FACE and other environmental groups (e.g. FIST – Friends Insist Stop Toxics) emphasised direct action strategies rather than negotiational tactics, with marches involving residents and media coverage. A local church sponsored a candlelight prayer vigil. Billboards were put up lining highways to condemn the proposal. Door-to-door canvassing was initiated. The state regulatory agency was sufficiently persuaded by these actions to refuse the permit to HIFI. In sum, the case illustrated how struggles to establish or reject a presence in the landscape were struggles between capital and community and strategies by capital to co-opt community members opposing landscape incursions failed. The case also reveals how a sense of place and attachment is a powerful base for resistance.

The riding of motorcycles has been a popular pastime since the 1920s. This form of recreation, particularly when undertaken in groups, quickly acquired a sub-cultural status, with distinctive sets of activities centred both materially and symbolically around motorbikes and a distinctive and identifiable 'uniform' (Willis, 1978). As a sub-group 'bikers' should be distinguished from the outlaw motorcycle gangs with highly structured rules and anti-social behaviour such as the Hell's Angels. Nonetheless in the eyes of the law all motorcycle riders are generally seen as potentially subversive (see Box 5.1). Willis (1978) recounts early instances of clashes between forces of the law, on the one hand, and biker groups, on the other, ranging from moral panics over the behaviour of motorcycle gangs on public holidays in Britain, to conflict over the wearing of helmets, regarded by the law as being an issue of public safety and by the bikers as being an issue of personal freedom.

For some, negotiational tactics and direct action such as picketing cannot succeed because of the images held of them in the first place, as rowdy and disruptive. The case of bikers (Box 5.1) illustrates how combined political, legal and cultural contestations of imposed cultural codes and police action are used as means of resistance, reflecting simultaneous adoption of symbolic and overt contestation. In a study of a battle centred on their freedom of access to public space, McDonald-Walker (1998) examines how motorcyclists contest images of and actions against them. She focuses on police actions in 1995 against meetings of motorcyclists at a public house in Warwickshire, Britain, called the Waterman and the motorcyclists' reactions to what they saw as unnecessary police harassment. The Waterman had, for about five years, been a landscape of 'bike nights' on Wednesday evenings, with the numbers particularly large in summer when bikers go for a ride and socialise with other bikers. As news of the evenings spread, more and more bikers came in the summer, some from as far afield as Liverpool, and the numbers swelled to 2–3000, according to police estimates. The Waterman could not cater for these large numbers. There were also many cars parked along the highway, belonging to those who came to look at the bikes. Further, people lined the roads to watch the bikes come and go, which may have led to local residents complaining. There were also a few among the bikers who behaved irresponsibly, 'pulling "wheelies" and "doughnuts" (spinning a motorcycle around a rotating front wheel)' (McDonald-Walker, 1998:380), and one biker who distributed leaflets promising a mini-racetrack and wheelie and doughnut competitions. That aside, the landlord of the Waterman insisted that the riders behaved well and there were no records of fights or arrests. Nevertheless, the police intervened and over two successive weeks, set up road blocks leading to the pub, preventing motorcyclists from gathering. Police were also at the Waterman as well as other local pubs. There was even a police helicopter and roving police motorcyclists.

Such overt conflict and contest over the bikers' use of public space was not well received. Bikers complained to the police about infringement of individual rights to travel and congregation and articles appeared in the local press

Box 5.1   Social conflict at the Bathurst bike riots

Cunneen and Lynch (1988) researched clashes between bikers and police in Bathurst, New South Wales, at the annual Easter bike races.

In the early 1960s, conflict centred on the park in central Bathurst which was symbolic hegemonic prestige space, complete with war memorial, fountains and formal gardens. The use of space was controlled by council by-laws. At that time numerous arrests were made for offences such as loitering, drinking alcohol, letting off firecrackers, swimming (in the fountain) and climbing (on statues). The bikers arrested over the years were predominantly working-class male youths.

By the early 1980s, the focus of conflict had shifted to the campground on Mount Panorama, which was public space. By that time the police had formed a crack response unit, the TRG, and in 1979 established a police compound on the mountain, surrounded by barbed wire. This compound was the focus of riots each year in the early 1980s.

Cunneen *et al.* (1989) undertook participant observation of the Easter riots of 1985. Their study found that the conflict, albeit real, was in fact institutionalised and formalised into a type of ritual game. The stimulus for the police activity on Easter Saturday 1985 was an evening of rowdy leisure activity, including bullrings (in which the crowd watches circular manoeuvres by bike riders leaving a doughnut-shaped impression in the ground), cockfights (fights with one biker atop another's shoulders) and games involving the throwing of lighted toilet rolls. By about 8 pm 'the last bullring had ended leaving about 1000 people, mostly young and overwhelmingly male, gathered outside the compound facing the police in a carnival mood' (Cunneen and Lynch, 1988:9). The scene was therefore set for the so-called 'riot', with ritual chants, throwing of missiles, retreat and advances (see Figure 5.1). Cunneen and Lynch reported that within five minutes a clear game framework had been set with the following moves:

- A squad of TRG in full riot gear faced the crowd.
- Crowd members, some 30 metres distant, threw rocks and bottles at the TRG.
- Every few minutes the TRG sprinted forward, sometimes tapping their shields with batons and making 'whoop-whoop' sounds as they ran.
- As a TRG charge began, the crowd would turn and run, usually in silence.
- A whistle blast stopped the TRG advance and they retreated, shields held high.
- As the TRG retreated, missile throwers stopped running, resumed missile throwing and ritual chanting such as 'pig, pig, pig' and other less polite terms.

The actions are similar to schoolyard games in which teams try to take territory from each other. However, in this case, it is doubtful either that the police wanted to take over the public space of the campground or that the crowd would have captured the territory of the police compound. Instead

(continued)

(continued)

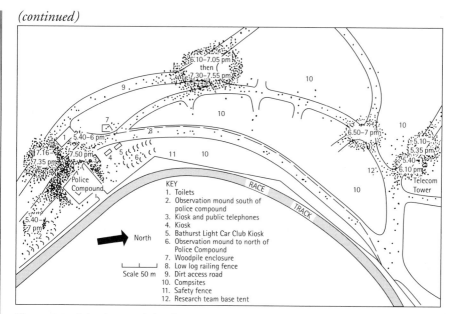

Figure 5.1 A landscape of ritualised conflict around the police compound, Bathurst race-track, Easter 1985

Source: Cunneen and Lynch, 1988, Leisure Studies, Vol 7, pp.1–19, Taylor & Francis Ltd., *www.tandf.co.uk/ journals.*

they were preoccupied with the contest itself. Some of the comments recorded from the crowd also reinforced the nature of the 'game' such as 'half time, change sides!' (Cunneen and Lynch, 1988:12).

Cunneen and Lynch see this ritualised landscape of conflict as being rooted in the class experiences of the biking community. It is an 'anti-police sentiment which momentarily resists and overturns the power structure while simultaneously reaffirming the 'absolute' nature of that power' (Cunneen and Lynch, 1988:17). As such the resistance, although real and heartfelt, is also symbolic and carnivalesque, its symbolic importance, on the one hand, affirmed by massive media coverage, its temporary nature, on the other, affirmed by the restraint shown in the limits imposed on physical damage. In another very real sense, the conflict outlined here is also a conflict over space. In the case of the park, appropriate use of war memorials and fountains required quiet and respectable behaviour and did not include drinking, climbing and other horseplay. In the case of the mountain campground, the police compound formed a spatial target focusing class-based and structural hostility between police and biker groups.

and radio as well as in Britain's largest weekly motorcycling newspaper. Many motorcyclists also wrote to their MPs. Possibilities for legal action against the police were explored although the action did not come to pass for various reasons: desire to protect the landlord; and bad legal advice which meant that it was past the time limit for registering an action even as funds were being raised to pay for a court case.

McDonald-Walker (1998) explored the reasons why bikers sought to contest the resistance that police put up against their presence at the Waterman and these reasons are instructive in revealing how cultural codes are being contested (Melucci, 1989). While the issue was about infringement of rights to the use of public space, to travel and congregation, it was also about motorcyclists' desire to challenge what they saw as outmoded representations of bikers as socially dysfunctional rebels. They consistently referred to themselves as 'responsible, productive members of society' (McDonald-Walker, 1998:384–5), thus emphasising their 'rootedness in everyday life and social roles' (McDonald-Walker, 1998:391) and arguing against negative stereotyping and prevalent unfair mythology. Their attempt at countering resistances against them therefore extended beyond the legal and political to the cultural, in which they attempted to gain control over social definitions of themselves. One revealing example is the way in which terms such as 'bikes' and 'bikers' are dropped in riders' rights literature for less emotive terms such as 'motorcyclists' and 'powered two-wheelers'. Lash (1990:19) argues that this avoids a certain 'invidious' valuation. Resistance to the 'cultural' may thus be as critical and powerful a strategy as legal and political contestation, if not more so. In brief, the resistances to the presence of bikers in the landscape and resistances by bikers to the attempt to eradicate them from the landscape illustrate both strategies and rituals of resistance.

A further example of resistances by a marginal group against powerful constructions of what is 'normal' and acceptable is the phenomenon described in some of the literature as 'gay ghettos'. The hegemony of heterosexuality proscribes non-heterosexual performance in most landscapes (see Chapter 4). Largely in response to this preclusion, gays and lesbians have developed their own spaces. Gay and lesbian precincts within inner-city areas have been labelled 'liberated zones' or pink or gay 'ghettos'. Castells' (1983) study of the Castro neighbourhood of San Francisco used the term 'liberated zone' to describe the array of gay venues and institutions in that area. However, he also refered to the area as the ghetto. Castells (1983:157) linked the development of 'the Castro ghetto' to the 'development of the gay community as a social movement' (see also Loyd and Rowntree, 1978:86). Similarly, the gay precinct in Jackson Heights, New York, was most overt during the gay rights campaigns when communities mobilised on issues surrounding safety, violence and homophobia (Almgren, 1994:52–3). Davis (1995:284) argued that the mobilisation of gay politics depended on such centres. These landscapes were fundamental to gay resistance against the hegemony of heterosexuality.

Gay precincts throughout the western world have developed a similar range of political and community support structures (Almgren, 1994; Castells, 1983; Hindle, 1994:20–1; Murphy and Watson, 1997:69–71). For example, the gay and lesbian organisations located around Darlinghurst, Sydney, have provided a community and political base. Community facilities include gay and/or lesbian night clubs, cinema and health clubs (see Figure 5.2). These

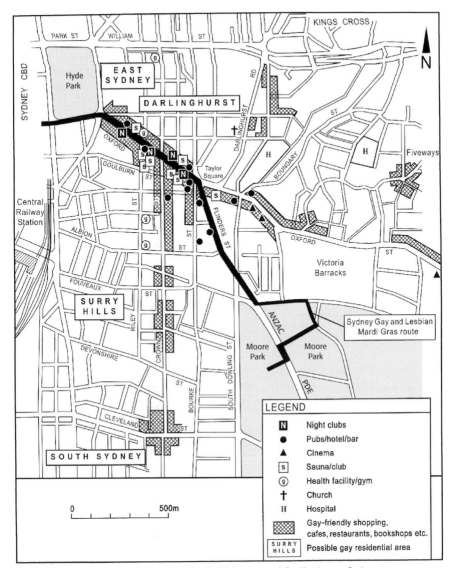

Figure 5.2   Gay and lesbian community facilities around Darlinghurst, Sydney
Source: Waitt *et al.*, 2000:479.

landscapes and events such as the annual Mardi Gras, which we discussed earlier, have been the source of political comment regarding HIV-AIDS, vilification, violence and official definitions of the family (Murphy and Watson, 1997:69). The lesbian and gay communities have been able to exert political influence on local representatives to the state and federal parliaments (Murphy and Watson, 1997:69–71). The South Sydney Local Government allocated spaces for organisations such as the Gay and Lesbian Rights Lobby, the Anti-Violence Project and the AIDS Quilt Project (Murphy and Watson, 1997:71).

The resistance and political mobilisation facilitated within these landscapes has clearly been substantial.

In the context of hegemonic heterosexuality and the violent prohibition of non-heterosexual performance (Chapter 4), lesbians and gays 'consciously seek out other sexual dissidents in "safe places"' (Valentine, 1995:96). The liberated zone, such as in Darlinghurst, Sydney's inner southern suburbs, is thought to serve as a site of safety for gays and lesbians. However, such precincts are also targets of homophobic violence and are foci for stigma and external scrutiny. Murphy and Watson (1997:83) observed that while the Darlinghurst precinct in Sydney 'bestows an identity and some degree of freedom', it is also 'a "gilded cage" where the prey is easily identified and picked off'. Data from survivor reports of homophobic attacks revealed that 44 per cent of all attacks in the state of NSW between 1991 and 1997 occurred in the Darlinghurst area. Almost half of all homophobic violence occurred on the street. What is more, 65 per cent of such street incidents occurred in the Darlinghurst area (Bennett, 1997:60). Bennett (1997:64–5) found that 'identifiable gay and lesbian sites, and streets where people felt comfortable "acting out" their identity, recorded high levels of violence'. A key informant from the Darlinghurst community commented that, 'you don't even have to look gay, like just the fact that you are on Oxford Street means that. People driving by in cars are going to yell out abuse at anybody' (informant, quoted in Bennett 1997:42). Clearly, the landscape is fundamental to the constructions of identity being weaved in these circumstances. As a focus of homophobic violence, the gay and lesbian landscape is more akin to a ghetto than a liberated zone. The resistance to heterosexual hegemony via landscapes is itself therefore not an uncontested phenomenon.

In yet another form of resistance against prevailing ideologies, there are those who 'work from within', as a kind of 'insider group'. In such a situation, instead of overt confrontation or direct challenge to cultural codes, groups avoid any kind of public outcry and seek to have their views heard and effected from behind the scenes. The work of nature conservation groups in Singapore illustrates this.

In a developmental state such as Singapore where the legitimacy of the state is built on its ability to promote and sustain development (understanding by 'development', the combination of steady high rates of economic growth and structural change in the productive system, both domestically and in its relationship to the international economy), conservation imperatives can quite easily present competing values. However, in a political culture where explosive and outright confrontation is absent, competing ideologies find other means of expression. In Singapore, the few existing environmental groups which seek to protect the limited pockets of nature have all operated primarily from 'insider' positions: they have generally not defined the agenda of environmental concerns as much as they have participated in the agendas set by the state. The Nature Society of Singapore (NSS), for example, is dedicated to the study, conservation and enjoyment of the natural heritage in Singapore, Malaysia and the surrounding southeast Asian region. The NSS was, until about 20 years ago, purely a nature appreciation group with no experience in environmental action. It was only in the early 1980s when the Serangoon estuary, Singapore's most bountiful bird habitat, was destroyed that the society awoke to the need to have its voice heard. From the start, the NSS operated primarily from an 'insider' position. In all their efforts, they did not seek to encourage public outrage; they consciously shied away from being labelled a 'pressure group'; and they preferred to engage in behind-the-scenes lobbying rather than open confrontation. In these ways, they succeeded in winning a certain degree of trust with the state, reflected for example, in the fact that the state accepted the NSS's proposal to conserve Sungei Buloh, a bird sanctuary, in 1988 and that the government has invited views from them on conservation matters. This has since encouraged the NSS to move a little closer towards setting its own environmental agenda, as reflected in its Masterplan on Conservation (submitted to the authorities in 1990), identifying areas of high ecological value to be preserved. To that extent, the society was successful in claiming legitimacy and the plan was endorsed by then Environment Minister, Dr Ahmad Mattar, who urged government departments and statutory boards to consider it seriously. To that extent, nature and the environment became explicitly acknowledged as Singapore's 'natural heritage' and the role of the NSS in making that agenda heard constituted part of the construction of a national ideology in which every citizen had a stake. In this way, another landscape was weaved into the Singaporean construction of national identity.

While resistance may be organised and strategic, oftentimes it is also diffused and unsystematic, with almost invariably poor results. Howdendyke, a village in Yorkshire, Britain, was designated an industrial estate in 1968 and, in 1979, industrial expansion was permitted if the industry were river related, given its location on the Ouse. Concomitantly, no new residential development was permitted. This plan would, in theory, support a stable residential village with some river-related industrial growth. However, in effect, the residential village was increasingly destroyed, with housing-related applications refused and demolition orders on about half of the residential dwellings issued. Existing housing deteriorated as repairs were not forthcoming. At the same time, new industrial sites were set up, with almost every industrial planning proposal approved since 1968. Residents sought to fight back, but the scales were uneven from the start. Resistance was ill organised. While some protested to the local Tory MP, others petitioned. Yet others tried to sell their houses, clear evidence that concerted collaboration in fighting the authorities was lacking. Topocide (the deliberate annihilation of place) was inevitable. Porteous (1988:91) sums it up thus:

> The conditions for the topocide of Howdendyke seem to be: a wholly working-class population, largely tenants, with little knowledge of the planning process and no tradition of dissent; control of land and employment by distant, profit-oriented corporations; planners willing to ignore their own guidelines and unwilling to engage in public participation processes; local politicians bent on development at almost any cost; and a certain deafness and lack of compassion on the part of all

these authorities. And topocide is even easier when the policies of planners, politicians, and industrialists dovetail so precisely.

For resistances to be successful and landscapes to be preserved, *ad hoc*, uncoordinated actions are seldom successful. Carefully crafted strategies for landscape protection on the part of communities are almost imperative in the face of strong ideological and practical agreement among powerful agencies: planners, politicians and capitalists. Landscapes depend on people and people depend on landscapes.

5.3.3 Balance of powers: intractable conflicts

Conceptually, not only do the powerful resist the incursions of the marginal and the less powerful contest the impositions of the dominant, it is also possible that situations exist in which two groups are locked in a balance of power, in which each seeks to exert its ideology and worldview, continuing in a protracted struggle for power and control, which may be expressed in territorial form. Nowhere is this more apparent than in the struggle between linguistic, religious and ethnic communities, which may be exemplified in struggles to secede. For example, linguistic communities, often although not invariably converging with shared religious, cultural and ethnic attributes, have resisted attempts to incorporate them within other ideological and cultural systems, and at times, have actively made claims towards self-determination, with ambitions towards territorial control.

In the contemporary world, one example is that of Sri Lanka, where the inhabitants of the northern and eastern provinces are united by the Tamil language, Tamil culture and a common history. These are bulwarks in a continuing struggle against Singhalese assimilation. Historians acknowledge that the Tamils have lived in these provinces since the thirteenth century. Under Portuguese and Dutch rule, these provinces were administered as a separate entity from the rest of the country. It was only under the British colonisers in 1833 that all parts of the country were under a unitary form of government, even while these provinces remained distinctively Tamil in constitution. This fact has been acknowledged in agreements at various historical moments: in the 1957 Bandaranaike–Chelvanayakam Pact, which recognised Tamil as a national language and agreed to the use of the Tamil language in administrative work in these provinces; and in the Senanayake–Chelvanayakam Pact of 1965 which pledged to make provisions for the use of Tamil in the northern and eastern provinces and promised to accord priority to Tamils in land allocation. United by their common aspiration for freedom and independence, the Tamil United Liberation Front ran in the 1977 general election and won a landslide victory. It called for a separate state. Later, the Liberation Tigers of Tamil Eelam (LTTE), or Tamil Tigers, emerged in the separatist fight and has come to be regarded by many as the representative of the Tamil population in the northern and eastern provinces. Insurrection, civil war and terrorism were still being waged at the turn of the millennium.

Figure 5.3   Republican murals in the city of Belfast, Northern Ireland

Another example of this is the case of Protestant and Catholic Ireland, in which the struggles have persisted for centuries and date back to the late sixteenth century. The city of Belfast illustrates a seemingly intractable conflict. In benign ways, the landscape markers of conflict include, within Protestant communities, the Union flag, the red, white and blue painting of street furniture and loyalist murals (see Figure 5.3). In Catholic communities, the Irish Tricolour may be found, the yellow, white and green colours and Republican murals. The effects of conflicts between the two communities, unfortunately, extend beyond such expressions, to real expressions of fear. Shirlow (1999) illustrated how the conflicts became manifest in how people behaved in various landscapes. For example, 75 per cent of adults in the Ardoyne (an area dominated by Catholic residents) would not be prepared to work somewhere if it involved travelling through 'enemy areas' (Shirlow, 1999:11). Indeed, job searching was considered to be influenced by fear for 68 per cent of Ardoyne respondents and 38 per cent of Upper Ardoyne respondents (Shirlow, 1999:10). The following newspaper account provides some indication of how this fear operated through space:

Cathy Bell, 34, a mother of three, is holed up in the working-class Protestant enclave of Glenbryn, ... She will not shop in the Ardoyne Road again, which traditionally served both Catholics and Protestants but instead, like an increasing number of her neighbours, she will go to another Protestant area, just as she takes

Megan [her five year old daughter] to a 'Protestant park'. She will rely on the mobile shop, marooned on wasteland in the middle of Glenbryn, for milk and cigarettes. And when she does venture out of this neighbourhood, she will go by car or taxi, too terrified to walk or take public transport through enemy territory (*The Independent*, January 2000).

Shirlow's team of researchers also noted that 'a quarter of the people surveyed in each study area have moved to their present homes either due to insecurity or intimidation while living in other areas' (Shirlow, 1999:7).

Fear and loathing of the Other in the Ardoyne reached international prominence in September of 2001, when some Catholic parents decided that they would take the direct route when escorting their children to the Holy Cross Primary School. Instead of taking a longer route, and entering the school through the back entrance, the parents and their small children walked along a 'Protestant street'. Loyalists rallied and stoned the parents and their children. These images of violence, and the subsequent conflict between Loyalists and security forces, were broadcast on CNN and in the major newspapers of the western world (see Mann, 2001:21). Shirlow (1999) concluded:

> This report conveys this sense of the localised nature of politics of territorial control, avoidance and resistance, where the imperatives of communal difference, segregation and exclusion have predominated over the politics of shared interests, integration, assimilation and consensus.

Clearly, in circumstances of cultural conflict, the cultural division of the landscape is a means for reinforcing apartness and exclusion (see Figure 5.4).

Figure 5.4 The Falls Road–Shankill divide, Belfast, Northern Ireland. War and fear of the 'Other' concretised in the landscape

Such cultural divisions are further manifest in landscape signatures which are reminders on a daily basis of cultural difference. Such landscape signatures can take the form of murals and graffiti (see also Chapter 3). Some of the most well-known landscapes of Northern Ireland are the political murals. The turn of the twentieth century was when murals first began to appear in Belfast (Jarman, 1998). The murals served as political statements for groups which had been banned by the British government from having their voices broadcast. Mural painting was illegal. For example, two men in 1970 were sent to prison for six months for painting an Irish Tricolour (Jarman, 1998).

One particularly interesting aspect of the murals is the focus on history (see Figure 5.5). Graham (1994) has referred to the murals, and other items of sectarian heritage, as spatial mnemonics. The murals are landscape reminders to the communities of why they should resist or hate. Jarman (1998) has argued that, in the 1990s, the murals have moved from simply being political statements. They have become landscape reminders to local residents of the conflict in Belfast. They reinforce senses of fear and have also become threats to those outside the community. Some murals are purposefully observable beyond the community boundary. Murals have also been resisted by some communities. Residents have painted over the artwork of the paramilitaries, only for it to be repainted in a cycle (Jarman, 1998).

In many senses the murals have become a symbol for Belfast. Television news journalists from around the world will have their story filmed in front of murals. Similarly, a photograph of a mural will often accompany international newspaper articles about Belfast or the Peace Process:

> The mural has become such a ubiquitous symbol of Northern Ireland. . . . Once the media had established that murals were the pre-eminent symbolic signifier of the northern conflict, the idea of Belfast could now be conjured up by little more than a few frames depicting a painting of a hooded gunman or King William on his white horse (Jarman, 1998).

In these instances, media texts are drawing on specific landscape features in order to generate constructions of society. By deploying the ubiquitous mural, social constructions of cultural apartness and conflict are reinforced.

5.4 Summary

In earlier chapters, we discussed how landscapes are both medium and outcome of power. In this chapter, we have shown how power is mediated and that landscapes also reflect and are reflective of conflict and resistance, not merely unmitigated power. Resistance may be the action of the powerful against the marginal or the less powerful against the dominant. It may be the continued tension between groups locked in struggle for supremacy or independence. Resistance may be symbolic and overt. It may be symbolically contestatory through the creation of extraordinary landscapes, but also through appropriating everyday landscapes. It may take the form of open defiance, direct action, legal action, negotiation or collaboration and change from within.

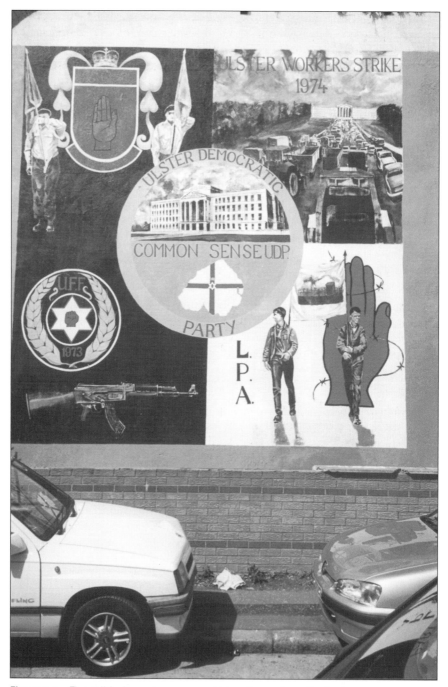

Figure 5.5   The political mural serves as a history lesson from the landscape

It may be organised and strategic or diffused and unsystematic. It may be a battle about racial ideology and it may be a contestation of gendered ideology. It may be a struggle for one's place, urban or rural, religious or secular, public or private. Like landscapes of power, landscapes of conflict and resistance are everywhere evident.

Chapter 6

Landscapes on the margin

6.1 Marginalisation and marginal landscapes _____

The photographs of Sydney, Australia, and that of the city of Manila in the Philippines show starkly different landscapes (see Figures 6.1 and 6.2). Metro Manila projects a corporate landscape typical of any CBD whose skyscrapers reflect the power of capital (see Chapter 4). Close by, the marginal district of Smoky Mountain is literally an enormous garbage dump picked over by its residents for waste materials. The city here smokes as the garbage heats, ferments and festers. People build their dwellings from waste materials and work ankle deep in mud and filth to extract useable and saleable materials. The polarity between the two is, however, not as clear-cut as the landscape contrast. The two parts of the city are linked by complex patterns of labour and material flow, the recycled goods and informal labour supporting the accumulation of capital.

These city landscapes are microcosms of global power structures that result in marginal landscapes occurring at a range of spatial scales. The processes of capital accumulation, industrial development and colonisation are central to understanding the spatial differentiation of many urban and regional landscapes. Landscapes of power and accumulation are significant symbolically yet may occupy only a small proportion of our cities (Chapter 4). In the same way, the proportion of capitalists, millionaires and people of power and wealth is very small within society. Most people are engaged directly or indirectly in the labour processes of production or reproduction, whether as factory workers, professionals, parents or children. Most landscapes too are drawn into this economic system, as sites of production and residence, such as the factory, the workplace and the house and suburb. In Chapters 4 and 5, some examples were discussed of these everyday landscapes, which reflect the values of the dominant hegemony for naturalised, common-sense reasons.

The focus of this chapter is on marginal landscapes, those defined as 'Other' to the powerful and the mainstream. Most suburbs and workplaces would not be defined as marginal. Some workplaces, such as the garbage dump of Smoky Mountain, the degraded landscapes resulting from mineral extraction, or the bleak social landscapes of the sweatshop, may be considered marginal. Although these landscapes may be integral to the success of the capitalist enterprise, they are sites stigmatised by various forms of oppression. Moreover, some suburbs,

128

Figure 6.1   The skyline of Sydney

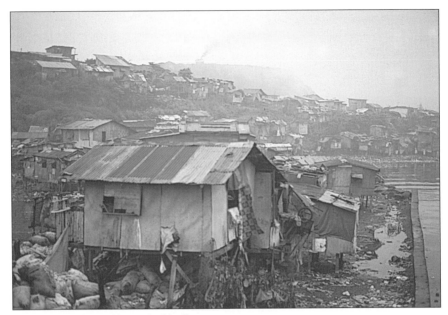

Figure 6.2a   Smoky Mountain, Manila

stigmatised by social problems, crime or racial tension, may also be categorised as marginal. They are inhabited by people who are outside or on the fringes of the system of power, an underclass, without jobs, housing, status or citizen-ship rights.

The processes of marginalisation have been defined by a number of writers, predominantly with reference to social groups rather than landscapes. None-theless the processes and categories are relevant to the landscapes they occupy. Winchester and White (1988) outlined three types of marginalisation (polit-ical, economic and social) resulting in a large number of marginalised social groups in the inner city such as refugees, ethnic migrants, the homeless and, in particular social contexts, single-parent families and gay and lesbian groups. In the case studies they employed, the city landscapes used by these groups varied, but included both stigmatised suburbs and public space used illicitly. Iris Marion Young (1990) defined five faces of oppression, each of which contributes to the exclusion of social groups and the landscapes it occupies. These five aspects of oppression were exploitation, marginalisation, powerless-ness, cultural imperialism and violence. In Young's work, the discussion of marginalisation is more specific than in the general sense in which it is used here; she uses the term primarily as a definition of economic marginalisation, specifically as an exclusion from the employment market. The landscapes of marginalisation discussed in this chapter are sites for all the oppressions iden-tified by Young.

It is important to emphasise that marginal landscapes occur in relation to all the forms and combinations of power identified in Chapter 4. The example

Figure 6.2b   Smoky Mountain, Manila

of Manila given earlier clearly has political and cultural dimensions as well as economic. The localisation of political or religious power brings with it spatial orderings of the landscape. Clear examples of such marginal landscapes occurred in the medieval cities of Europe, where Jews, lepers and the excommunicated were cast literally 'beyond the pale', i.e. beyond the city walls (White, 1984). In the early years of the twentieth century, the stigmatised Jewish ghettoes of the cities of eastern Europe were an easily defined target for racially motivated pogroms which decimated the Jewish communities and forced their widespread diaspora. Similarly marginalised landscapes may occur in relation to cultural norms, such as the norm of heterosexuality. The classic examples of gay areas of cities have been examined in Chapter 4. In some ways these marginal spaces have become mainstream, renovated by gentrification and supported by the tourist dollar (see also Chapter 5's discussion of the Mardi Gras parade); nonetheless they still challenge the hegemony of heteronormativity. Lesbian landscapes as landscapes of personal transformation are considered in a later part of this chapter.

The chapter considers first those landscapes that have become obsolete because of the changing nature of the capitalist enterprise and which have been abandoned and in some cases transformed. The transformation of old industrial landscapes into heritage sites and theme parks involves a shift in symbolic meaning and a privileging of particular aspects of history. These are landscapes that are predominantly economically marginal.

The second focus of the chapter is on landscapes of the poor. Many of these landscapes are transitory and may be invisible to, but co-existent with, the mainstream city. In a number of cases, these are landscapes that do not conform to the dominant constructions of race, gender and sexuality as well as capital and class. These are landscapes that are economically and socially marginal, often where the groups are subject to cultural imperialism.

Finally liminal landscapes are considered in this chapter, landscapes which in one way or another are 'on the edge' (Shields, 1991) where meanings and social categories can be transformed, as in Rio's Carnival or sites of pilgrimage. Such landscapes may be at times sites of conflict and resistance (see, for example the discussion in Chapter 5 of carnival). Liminal landscapes are different from the marginal landscapes just defined as 'Other' to the hegemonic. In many cases, such as the famous example of Niagara Falls, they are integral to the dominant system, in this instance, as a major tourist attraction. Nonetheless, they have become imbued with symbolic meaning as places where personal transformations can occur (Shields, 1991).

In a number of cases, the distinctions that we have drawn between, for example, landscapes on the margins and landscapes of the everyday, or between marginal landscapes and contested landscapes, are less clear in reality than in this text. So, for example, the large public housing estates on the edges of cities grade from being integral to the capitalist system to being margins of social disorder; the marginal landscape of the Gypsy encampment is also the site of conflict, expressed physically, administratively and in discourses of difference. In such cases, we have placed the examples where we have considered them to be most appropriate and have cross-referenced them where possible to other related aspects of these landscapes.

6.2 Obsolete landscapes

Many regions founded on old industrial capitalism were built on the heavy industrial complex of coal, steelmaking and heavy engineering. The closure of these industries, as a consequence of outdated technology and international competition, has altered their landscapes. Formerly a hive of productive activity, such landscapes have become progressively derelict, scarred by coal mining, poisoned by industrial waste and littered with abandoned factory sites. Such landscapes have formed problem regions such as the Ruhr in Germany, the Midlands of Scotland around Glasgow and the Pennsylvanian coalfield around Pittsburgh. Industry became a dirty word, conjuring up images of blackened smokestacks and unemployment queues rather than positive images of productive activity and gainful employment. Short *et al.* (1993), in their discussion of Syracuse, commented that the connotations of industry had become exclusively negative. This view particularly contrasts with the 'clean and green' image projected by newer high technology activities, which process knowledge, information, transactions and finances rather than the physically messy activities of processing raw materials and manufacturing them into products.

Similarly, dockland areas of formerly congested wharves made obsolete by the huge increase in the capacity of ships and by computerised bulk loading technology have been transformed. Docklands and freight facilities have become waterside playgrounds, incorporating festival market places geared for consumption activities with dining areas, entertainment and leisure facilities and specialist shopping (see the discussion of Darling Harbour, Chapter 4). Adjacent warehouses and stores have been turned into gentrified apartments and units, complete with security services, gyms, pools and health clubs and coffee shops aimed at a professional and well-heeled clientele.

These transformations of landscape do not just happen. In many cases, the activities of boosterist city councils and local governments have been primarily responsible. Landscape changes have been brought about as a consequence of reinvestment decisions made to make the city more competitive in attracting mobile capital. These are often accompanied by powerful legitimating discourses which emphasise the timeliness, appropriateness and inevitability of the change. In the case of Vancouver, the council was elected on the platform of a 'liveable city', which had the aim of improving and making functional the obsolete landscapes. The unintended result of this upgrading was massive gentrification and displacement as a consequence of increased land values (Cybriwsky *et al.*, 1986; Ley, 1981). In Society Hill, Philadelphia, a coalition of developers and councillors, whose interests were not clearly defined, combined to introduce reinvestment on a massive scale to a devalued industrial/commercial core (Smith, 1979). The redevelopment of Darling Harbour in Sydney required government funding and special legislative powers to create a

'Bicentennial gift to the nation' (Huxley and Kerkin, 1988; see also Chapter 4). In each case, a state or government initiative has been followed by international and individual capital responding to perceived opportunities for consumption of goods and new residential lifestyles. In the case of Homebush Bay, the Sydney Olympic Games provided the necessary impetus for the substantial rehabilitation necessary (see Box 6.1 and Figure 6.3).

Box 6.1   Homebush Bay: from toxic dump to international stage

The principal site for the Sydney 2000 Olympic Games was an area referred to as Homebush Bay, which is located off the Parramatta River which flows east into Sydney Harbour. These sporting venues are in central western Sydney, a generally poor area with high numbers of residents with only basic English language skills, low average incomes and a weak rates base (Owen, 2002). The preparation of the sporting venues at Homebush Bay, and the development of the adjacent agricultural showgrounds, involved a large-scale rehabilitation of what was a toxic wasteland. The prior land uses of the Homebush Bay area included chemical manufacturers, brickworks and abattoirs. The Olympic Village was constructed on top of a landfill garbage tip and an Australian Navy ammunition dump (Figure 6.3). The area was, and perhaps still is, badly contaminated by industrial and household wastes.

The restructuring of industrial places has been found usually to involve a change to the physical fabric and a refashioning of identity (Dunn et al., 1995; Short et al., 1993). A transformation of the fabric of the city is required in order to remove the material industrial legacy. The emerging landscape must accommodate the anticipated new economic base. The urban landscape must be made appropriate to post-industrial activities, such as tourism, education or recreation. Homebush Bay in western Sydney has undergone a profound physical and symbolic restructuring.

Olympic Games planning in Sydney involved a great deal of public monies, A$2.3 billion as of early 1999 (Dunn and McGuirk, 1999). This substantial investment has had some significant impacts on the landscape of Homebush Bay. It is difficult to imagine how the physical problems in Homebush Bay could have even begun to be addressed if it were not for preparations of the Olympic hallmark event. Rehabilitation of this site was beyond the means and responsibility of the local authority, Auburn City Council.

Not only were noxious, if not toxic, sites physically rehabilitated, but the imagery of Homebush was altered. The landscape has shifted from being seen as a dirty, polluted industrial zone, to a post-industrial, clean and green place of leisure and sport (Dunn and McGuirk, 1999). There is little doubt that Homebush Bay has already become associated with the 2000 Olympic Games, as well as the smaller hallmark events such as sporting finals and the annual agricultural shows. This marks a significant transition from its previous identity as a noxious industrial dumping area.

Figure 6.3   Homebush Bay, principal site for the Sydney 2000 Olympic Games. Location of the Athletes' Village, in its previous guise as a Navy ammunition store

The financial reinvestment to turn around obsolete landscapes has not occurred in isolation. Inevitably, the change in landscape is accompanied by a change in discourse. Industrial places have been reimaged, repackaged and marketed as desirable commodities for consumption. This marketing of place often constitutes a major break with the industrial past, as themes and events are used to manufacture a new present. The repackaging of landscape and a change in discourse is markedly evident in the city of Newcastle in Australia. The phases of landscape development have been outlined in Chapter 2 (Box 2.4). In the reconstruction of place, some histories are privileged while others are expunged from the collective memory. Newcastle's promotional material has, until very recently, almost completely failed to recognise both its Aboriginal and its convict heritage. Lately, however, there has been a rash of public art commemorating both these eras, as a token of the importance of reconciliation between white and black Australians. The increasing distance from this uncomfortable past also allows a romanticisation of the appalling events of dispossession and brutality.

Repackaging of the Australian city of Newcastle also ignored the industrial past, industrial landscapes and the manual labour on which they were built. City representations emphasise images of families enjoying the beach, couples sipping wine on the waterfront and women admiring the sporting achievements of their menfolk. In these representations, the dirt of industry, the labour of the working class and the role of women are noticeably invisible (Dunn *et al.*, 1995). In the process of place marketing, history and landscape are commodified into saleable chunks. At present, 'industry' still has connotations of dirt and exploitation and the industrial character of Newcastle's landscape is too recent and too raw to be preserved as heritage. Some small portions of the built environment will be preserved as sanitised heritage items and museums, but the majority will be bulldozed and the heritage value concreted over with the industrial contaminants, probably to provide locations for a new generation of leaner, greener industry. In this case, as with many others, the problem of technological obsolescence is compounded by physical contamination and by the social construction that is placed on the industrial past. Industry is seen as a constraint preventing the city from moving forward and cannot yet be preserved or celebrated until those shackles have been removed. Ways of dealing with obsolete landscapes reflect partial but dominant constructions of both the city as it was or might have been and particular visions and values for the future.

In Singapore's Chinatown, divergent views of the past and what constitute desired landscapes to replace obsolete ones are evident (Kong and Yeoh, 2002; Yeoh and Kong, 1994). Post-World War Two Chinatown was an area 'occupied by acre after grim acre of ramshackle shophouses in which gross overcrowding was common' (*Homes for the People*, no date:30). Severe residential overcrowding was conflated with other problems: insanitary buildings; high land values and irregular plot sizes; the lack of public open spaces and community services; a complicated mix of residential, industrial and commercial land use in close proximity; and road congestion (*Master Plan Report of Survey 1955*). With independence in the early 1960s, as part of the state's bid to secure political legitimacy, to build ideological consensus and a disciplined industrial workforce (Chua, 1991), the eradication of housing difficulties and slum problems in areas such as Chinatown gained prominence on the urban agenda. These landscapes quickly became obsolete, deserving removal and replacement. The transformation of the original Chinatown landscape was not only perceived as a means of improving living conditions for the people but as both prerequisite for and tangible proof of larger forces of socio-economic development and progress at work in the state. The hitherto ubiquitous shophouses were also perceived as constituting an uneconomical use of valuable urban land and as having 'quite clearly outlived their purpose' (URA, 1983:7). New house forms consistent with the aspirations of a modern city were needed. Thus, from the state's perspective, the 'Herculean task' of slum eradication and its replacement by low cost public housing and medium cost private housing represented 'a stake in the future, in stability, prosperity' (HDB, 1969:59). Singapore's ambition was to be able to take pride of place in becoming 'an integrated modern city centre worthy of Singapore's present and future role as the "New York of Malaysia"' (*Homes for the People*, no date:84).

In less than two decades, the built environment of Chinatown was dramatically redrawn along modernist lines informed by efficiency, discipline and rationality of land use. In the early 1960s, it was estimated that a quarter of a million people required rehousing if Chinatown were to be redeveloped. By the mid-1970s, the basic fabric of old Chinatown incorporated many new elements. Demolition went in tandem with building 'homes for the people' to accommodate families which had to relocate. Some public housing in the form of high-rise low cost flats which allowed a rational use of high-value centrally located land was provided within Chinatown itself in order to keep the population in the city centre.

Despite the rhetoric of progress in which urban renewal was couched and the removal of obsolete landscapes anchored, the rewriting of the Chinatown palimpsest did not entail the total erasure of the amalgam of forms laid down during the pre-independence era. While certain parts of Chinatown did not escape the bludgeon of redevelopment, sufficient vestiges of the shophouse motif endured. For many years, the fate of the remaining old Chinatown landscape stood in the balance, but by the late 1970s, there were signs of a rethinking of the overall state policy pertaining to Chinatown. While this did not signify an overturning of the redevelopment juggernaut, it was symptomatic of the wider concern that transforming historically significant and culturally rich landscapes into an 'environment of towers' would dilute the country's heritage, an ingredient crucial to the pressing task of nation building (*HDB Annual Report*, 1961:4). Investigations into the viability of conserving Chinatown were set into motion as early as 1976/77, but these efforts did not come to fruition until 1986 when the Urban Redevelopment Authority (URA) announced its Conservation Master Plan, which included the preservation of substantial portions of the Chinatown landscape. In the meantime, during the hiatus, serried rows of two- and three-storeyed shophouses with their narrow staircases and tiny cubicles continued to form a principal part of the Chinatown landscape, belying the portrait of a modern, technologically progressive city, an image fundamental to the logic of urban renewal. State regulation of the remaining shophouse landscape harked back to an earlier era, consisting essentially of by-law control, sanitarianism and surveillance. The housing landscape of 1980s Chinatown was hence one of contradictions: beneath a skyline punctuated with tower blocks and high-rise buildings, an undulating sea of 'obsolete' shophouses continued to persist.

In everyday usage, the housing landscape of Chinatown of the early 1980s was one inscribed with contradictory and diverse meanings. In spite of the frenetic pace of change in the area, for those left in the remaining shophouses, living conditions remained fairly stagnant with few significant breaks from the pattern which had prevailed since the colonial era. Despite increasingly more stringent regulatory measures aimed at sanitary improvements since independence, many families and individuals continued to live out their lives in crowded, windowless cubicles of plank and cardboard rented to them by chief tenants; share communal kitchen, bathroom and toilet facilities, depend on kerosene lamps for lighting; tolerate a general state of disrepair; and endure the hazards of fire where the only access to each floor was via a steep and narrow wooden staircase located in the front part of the house. While the presence of refrigerators, television sets and telephones indicate that residents' lives were not totally untouched by modernisation, the 1980s shophouse landscape in Chinatown represented pockets where housing conditions had essentially fossilised, where residents experienced little of the state's efforts to upgrade physical living conditions.

Modern living, much vaunted as part of the state's march towards progress and an integral strategy in the eradication of obsolete landscapes, is hence a mediated experience in the Chinatown housing landscape. 'Obsolescence', however, became overturned as the late 1980s and 1990s developed into a time of reflection about cultural roots and a period of increased reliance on the tourist dollar. In the late 1980s a new conservation ethos which placed the accent on reclaiming heritage and capitalising on the city's 'more traditional assets' emerged (*URA Annual Report*, 1986/87:2). As part of this conservation effort, the state capitalised on what it deemed to be positive Chinese cultural traits. State agencies such as the Urban Redevelopment Authority (URA) and the Singapore Tourism Board (STB) have identified Chinatown with the pioneering spirit and enterprise of early Chinese immigrants to Singapore and showcased it as a distinctively Chinese cultural area which 'brims over with life, capturing the essence of the old Chinese lifestyle in its temples and shophouses and nurturing a handful of traditional trades [such as] herbalists, temple idol carvers, calligraphers and effigy makers . . . in the face of progress' (*Singapore: Official Guide*, 1991:28–9). Against a backcloth of shophouses and temples, large-scale festival activities, fairs, *wayangs*, puppetry and trishaw rides were to be 'staged' to provide both locals and tourists with 'a different kind of experience' (*Conservation within the Central Area*, 1985:15). Particularly during Chinese festivals, lion and dragon dances are brought in; national Chinese calligraphy competitions and exhibitions are held; ancient Chinese lantern quizzes are hosted; and Cantonese operas are performed (*The Straits Times*, 19 February 1985). In conserving Chinatown as a testimony to the vibrance of Chinese culture, state strategies focus in the main on refurbishing the traditional architectural facade of Chinatown buildings. Where market forces permit, the resurrection of 'vanishing trades' which are perceived to epitomise the Chinese past is also encouraged. Conserved Chinatown is conceived as a quintessential Chinese hearth, both for the tourist gaze as well as for locals in search of the vanished past. What had earlier been construed as obsolete and marginal had come to be inscribed with symbolic centrality.

The conservation of Chinatown, both in terms of physical facade and activities, reflects attempts to anchor a particular version of Chinese identity in order that Singaporeans may find cultural anchor and rootedness in a world besieged by western influences. Yet, at the same time, in the attempt to reclaim Chineseness, the state is also careful to tread the thin line between recreating ethnic identity and embracing the larger 'nation'. Indeed, the state seeks to elevate Chinatown to national importance as a civic asset, 'a common bond place' for 'Singaporeans living in outlying new towns' (*Conservation in the Central Area*, 1985:15). Conserving Chinatown as a veritable repository of tradition, history and culture can thus be understood as an attempt to render heritage in material form and hence serve the sociopolitical purpose of binding Singaporeans to place, to the city and, ultimately and vicariously, to the 'nation'.

While conservation has fostered a state-envisioned 'Chineseness' embodied in distinctive architecture, a scattering of unique 'dying trades' to represent the Chinese past and a variety of 'Chinese' festive activities, it has also led to the demise of much more prosaic elements which go into the making of a Chinese lived culture. Shopkeepers, families and street vendors lament the rapid attrition of longstanding small businesses which have been part and parcel of the familiar landscape where the retailer–client relationship goes back a long way. This is contrasted to the cautious way in which they view the sudden influx of gentrified shops managed by new people. They are far from persuaded that what the URA promotes as 'adaptive reuse' of traditional buildings has revitalised the 'traditional Chinese way of life'. The Chinese in Chinatown have remained aloof to such a version of the landscape and are not entirely oblivious to the irony that in its attempt to refurbish Chinese architecture and revive Chinese 'dying trades', the state's conservation efforts have essentially damaged the day-to-day cultural life of the place, creating perhaps a different kind of obsolete landscape.

Particular examples of obsolete landscapes in urban areas have been dignified by the acronym of TOADS (temporary obsolete abandoned derelict sites) in a number of cities on the eastern seaboard of the United States (Greenberg *et al.*, 1992). Deserted commercial and industrial sites and derelict and abandoned housing properties became the centres of health, social and environmental problems including toxic waste dumps, crack houses for drugs, arson, crime and squats for the homeless. In the Bronx, it was found that the TOADS were associated with a rising incidence of TB, homicides, infant mortality and drug-related AIDS. Greenberg and his co-authors found that TOADS were most likely to occur in city neighbourhoods with a weak real estate market, a high proportion of poor, minority residents and a diminishing economic base. They noted, however, that not all districts with these economic and demographic characteristics exhibited such problems. In particular, areas with active minority communities and/or effective elected officials were seen to be both less susceptible to the development of TOADS and more successful in their amelioration (Greenberg *et al.*, 1992:120).

In an environment of declining federal support programmes, the prime responsibility for amelioration lay with the local government. The authors suggested that efficient and successful ways of dealing with such TOADS included effective monitoring of land uses, the maintenance or improvement of fire and emergency services, the provision of controlled dumping sites and the conversion of TOADS into viable activities such as homes for the homeless. Often such solutions are introduced through the work of the elected officials and the strength of the minority community.

Similar means have been successful in San Francisco's Tenderloin district where the non-profit Community Development Corporation (CDC) has provided a link between community, capital and government in providing sensitive neighbourhood regeneration (Robinson, 1996). Two of its most successful ventures have been the successful redevelopment of the Cadillac Hotel as non-profit housing for ex-convicts and the Senator Hotel as housing for the homeless. In both cases, the buildings had experienced severe disinvestment, were predominantly uninhabitable and the scene of drug dealing and crime. The CDC, although not feasible everywhere, is considered by Robinson

(1996:1663) to be 'weaving a sense of community coherence, meaning and value' in the regeneration of marginal landscapes in inner cities.

TOADS, as areas of economic obsolescence, became increasingly occupied by marginalised and oppressed groups as their economic base declined. The physical descriptions of the TOADS given by Greenberg *et al.* (1992) summarise a discourse of decline in an almost mythical recital of the dangers of the inner city (Burgess, 1985), which is discussed in more detail in the next section. Similarly, the recipes for successful regeneration are clothed in evocations of hope, based on notions of personal and community responsibility. With the removal of the economic base and the abdication of responsibility by higher orders of government, the marginal landscapes provided spaces for empowerment at the individual and community level.

6.3 Landscapes of the poor

Soja (1993; 1995), in writing about the postmodern city, characterised it in part in terms of multiple polarisations. The increasing polarisations between social groups create and reinforce socio-spatial boundaries and are at the heart of marginal landscapes. Enforcement of exclusion, he argues, occurs increasingly through design measures, the introduction of by-laws and ordnances, active surveillance, control and policing and, importantly, by social constructions of marginal groups through discourse, stereotyping and moral panics. In this section, we introduce you to different landscapes of the poor, including those marginalised by both economic forces and by social construction. In examining marginal landscapes such as service-dependent ghettos and ethnic minority ghettos, nineteenth-century workhouses and landscapes of Gypsies and hippies, we illustrate how economic forces and social constructions may reinforce each other in enhancing polarisations.

Michael Dear and Jennifer Wolch (1987) have written chillingly about what they have termed 'landscapes of despair' in cities in Canada and the USA. These are neighbourhoods adjacent to the city centre that have accumulated numerous facilities for the service-dependent poor. The services include boarding houses and group homes, out-patient services and drop-in clinics catering for the homeless and de-institutionalised. These zones of service provision tend to develop in areas which have many of the characteristics of those described earlier as being susceptible to TOADS development. They are areas of mixed land use where zoning allows group rather than single-family homes and institutional rather than residential uses. Their land values are diminished by obsolete industrial land and decaying tenement buildings. Their communities are likely to be poor, marginal, fragmented and disempowered.

Proposals to locate group homes for the de-institutionalised mentally ill or an AIDS hostel for the terminally sick meet concerted opposition in more affluent and established suburbs, where they are prohibited by zoning laws, priced out of the market by more affluent bidders and confronted by organised resident action groups. The location of such facilities is therefore predominantly affected by external factors and, as a consequence, they cluster where resistance

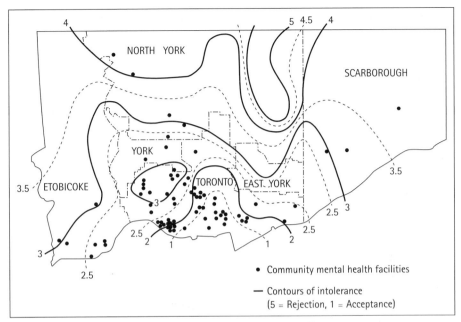

Figure 6.4 The contours of intolerance and social exclusion, Toronto
Source: Dear and Wolch, 1987:107.

is lowest (Hall and Joseph, 1988:305). The net result is a concentration of the service-dependent poor in deprived areas of the inner city; a landscape of despair, a service-dependent ghetto.

The development of service-dependent ghettoes has been studied in depth in the cities of Toronto and Hamilton in Canada (Dear and Wolch, 1987). In particular, the location of group homes, containing many of the de-institutionalised mentally ill, became concentrated in neighbourhoods close to the main railway station and CBD. Many of the people who were de-institutionalised required constant access to day-care facilities and so located in areas close to such services. This pattern of dependence brought about what Dear and Wolch termed a 'drift to the (service-dependent) ghetto' (Dear and Wolch, 1987:133) as well as a distinctive physical and social landscape. The isolines that Dear and Wolch drew on their map of Toronto reflected what they considered to be the contours of intolerance and the realities of social exclusion (see Figure 6.4).

These landscapes of despair are brought about through the operation of market forces. First, the city centre has undergone disinvestment as industry has contracted and the wealthy have filtered out to the suburbs. Second, the supposedly rational and efficient way to provide services for groups such as the mentally ill has been to provide services 'in the community' rather than in institutions. Third, the planning control and development principles have served the interests of the affluent and articulate in preserving the better suburbs from incursions by people who are perceived to be socially undesirable. Landscapes of the inner city are polarised between the extremes of rich and poor. It has been argued consistently by a number of authors that the polarities are getting wider and the boundaries or collision zones between them more marked (Soja, 1993; Winchester and White, 1988). The contrast between the corporate CBD or the consumerist condominium, on the one hand, and the run-down infrastructure of the service-dependent ghetto, on the other, could not be more marked or more visible. As in the case of Manila cited earlier in this chapter, these landscape differences can misleadingly suggest separation, yet they are linked by circuits of capital.

However, in the case of the service-dependent ghetto, as in Manila and the inner-city TOADS, there are not only market forces at work, but also forces of power and social construction revealed through discourse. Those in power have had the capacity to manipulate government systems to protect their own investments and to exclude people and land uses deemed to be undesirable. The discourses used in opposition have constructed the service dependent as stigmatised and morally dangerous people who will confront and corrupt 'respectable' residents in 'normal' residential areas. Narratives of dirt, danger and perversion are deployed as tactics of urban exclusion. Furthermore, the stigma attached to particular landscapes, such as the sites of former lunatic asylums, is a powerful one which is more easily reinforced than overturned.

The landscapes of the poor literally get a bad press. Powell (1993) found in her analysis of press treatment of the Mount Druitt suburb of Sydney, Australia, that scuffles between youths were enough to taint the whole of western Sydney as a wasteland of incivility, disadvantage and danger. The media are instrumental in maintaining myths of landscapes of the poor. Myths, like social constructions, have elements of truth but transform a particular history into a naturalised social order (Barthes, 1972:143). The role of myth has already been examined in the example of the city of Venice (Chapter 4). The lived reality of the landscapes of the poor is subsumed in myth, which Burgess (1985:206–7), writing in the context of the UK, considered to have four elements; the physical environment of 'derelict houses and tatty streets' with its common-sense link to violence and vandalism; a white working-class culture, consisting of a combination of unemployment, low educational achievement, booze, sex, inadequate housing and poor parenting; a marginalised black culture, in this case predominantly immigrant West Indian, characterised by alienation, youth crime and low levels of educational attainment; and a street culture, consisting of prostitution, vandalism, mugging and graffiti, in a combination of illegality and loose moral standards.

Beyond economic forces, therefore, social constructions of landscapes of the poor further marginalise and stereotype their inhabitants. It is notable that Burgess' (1985) description of the physical landscape draws on the imagery of the nineteenth-century industrial city, with the negative connotations common to industrial landscapes. The myth is also critical of post-war redevelopment and public housing, while casting both black and white working-class communities as alien to hegemonic white bourgeois culture. Burgess (1985:223) concludes that: 'The myth removes the places and the people who live in them to a grey, shabby, derelict, poverty-ridden fairytale-land which can be conveniently ignored because it has no reality.' In the case of British inner cities, they can be conveniently mythologised and thereby ignored, until the next episode of 'race rioting' brings them back to public attention, as happened in the riots of 2001.

Immigrant or minority ethnic groups occupy many of the most marginal areas of cities. Areas designated as ethnic ghettoes are often imbued with pervasive racially based stereotypes. In one of the first examinations of this, Burgess (1985) described how the West Indian community in Liverpool was stereotyped and cast as an invading alien force. The media have long been recognised as central to the social construction of ethnic minorities as problems. Geographers have demonstrated how stereotypes are reinforced through the social constructions of the spaces of minority groups (see Dunn and Mahtani, 2001).

Media negativity is uneven across cultural groups. For example, Muslims and Asian–Australians in Sydney experience a much greater degree of media negativity than do other community groups. The impacts of Islamophobia in Sydney were discussed in Chapter 5. Anti-Asian sentiment has been focused on the Indo-Chinese communities in the outer western suburb of Cabramatta. The media treatment of Cabramatta has been one sided and inflammatory. The media sensationally treat Cabramatta as a site of crime and poverty (Dunn, 1993:229). By focusing on issues of violence, crime, vice and policing in Cabramatta, the media reinforce orientalist social constructions of Asian–Australians as criminal and depraved (Dunn and Roberts, forthcoming).

The maintenance of some poor inner-city areas undoubtedly serves the interests of capital and the state by keeping poor and marginal activities away from prime space while providing a reserve of labour for current exploitation and land for future development. Many areas of longstanding poverty and dilapidation are therefore of interest to developers, particularly if they have benefits of accessibility to the city centre as well as property that is ripe for gentrification or redevelopment. Some examples of such inner-city revalorisation and reimaging have already been discussed in this chapter, such as in Singapore and Newcastle, Australia.

Marginal areas of public housing built on the edges of cities are much less likely to come under redevelopment pressures. They may nonetheless be even more significant as areas of deprivation because of their size and scale of development, the numbers of residents and their distance from transport, services and employment. In the UK, the extent of marginal council estates is huge: approximately five million people are housed in them. There is clear evidence that poverty in these estates has been intensifying through the 1980s and 1990s (Power, 1996). The country was alarmed when the early 1990s saw riots of the type which had occurred in Toxteth, Liverpool (Burgess, 1985) and which had been thought to be more typical of black inner-city tenement areas. The occurrence of the riots provided incontrovertible evidence of severe alienation, particularly of young males. The decline of social programmes and welfare services, the rise in crime, the underfunding of schools and the desperate need for employment compounded the malaise.

In Australia, Diane Powell (1993) has graphically summarised the stigma of some of the large estates built on the western and south-western edges of Sydney. These estates contain the highest proportions of 'problem' families and many are poorly serviced and inaccessible. In the 1990s the public housing estate of Claymore in Campbelltown had no bank, poor public transport, shops encrusted with graffiti and guarded by steel grilles, significant levels of unemployment and the highest proportion of single-parent families in Australia. On a smaller scale, the northern Wollongong suburb of Bellambi in Australia is a stigmatised public housing estate with similar although less concentrated socio-economic characteristics (Winchester, 1992). In estates such as these, single-parent families (and especially lone mothers) have been demonised as destroyers of the social order based on the patriarchal family (Winchester, 1990). In fact, most of the single mothers result from divorce after unsuccessful and in some cases, stressful and abusive marriages. As a result of divorce and childcare needs, many single parents (who are mainly mothers) are living in poverty and are also potential victims of violence and property attack in their homes. The feminisation of poverty has become concentrated through public housing policies which have now been replaced by a policy of dispersion. In these suburbs, however, the landscape of stigma is reinforced by the social construction of single parenthood. The example of Bellambi is an interesting one as the estate is built adjacent to the beautiful surfing beach of Bellambi, on land which is changing in value from being a straggling coastal settlement to what is becoming prime real estate. In the last few years, the wetlands and dunes on its northern edges have been developed as prestigious suburban homes. It will probably only be a matter of time, and changing discourse, before the landscape of stigma is creatively destroyed to make way for capital investment with a higher financial return.

Within and around the prime real estate of the city are other marginal spaces that are less visible and more transitory than the residential areas with which the last section has been concerned. Many marginal groups, such as the homeless, prostitutes (see Chapter 5) and street kids require the use of prime space for their essential activities: the homeless for begging or 'panhandling', street kids for shock and display as well as for other purposes and prostitutes to engage with their clientele. In Paris, street kids and punks would gather in major public spaces such as the Forum-Les Halles where the major subway intersection and underground shopping centre combined to provide great accessibility, exposure to the public and shelter from the elements (Winchester and White, 1988:49–50). The places used by these young people, as well as the places used by buskers and beggars, are prime spaces but being used subversively for marginal purposes. Similarly, zones of street prostitution make use of public space, although often severely scrutinised and policed. These have been considered in greater detail in the previous chapter, while the marginalised landscape of the Filipino sex industry is the subject of Box 6.2.

One group which is already on the edge are urban Gypsies. Sibley's (1999) study of Gypsies as 'outsiders in society and space' accurately portrays their social and spatial position. Gypsies are an ethnically and linguistically defined

---

Box 6.2   'Cartographies of desire': marginal/ised prostitutes in the Philippines

As Law (2000) pointed out, the sex industry and its workers can be constructed in different ways in different contexts. In the vast majority of such constructions, the prostitutes are a marginal/ised group in many cities. They are contained both literally and geographically within 'red light districts'. In one construction, involving health professionals, they are a marginal group requiring epidemiological focus, concerned as health professionals are with the spread of sexually transmitted diseases, including AIDS. In another construction, establishments of prostitution are viewed as 'inscriptions of foreign control in the landscape, enabled by government policies on tourism' (Law, 2000:29). The following example illustrates these different constructions.

In Cebu, the Philippines, for example, Law (2000:32) illustrated how 'the emergence of bikini bars catering to white, Western men coincided with Cebu's emergence as a cosmopolitan city with Western service outlets'. At the same time, the sex industry there developed further with the increased number of American visitors during the Vietnam war as Cebu's airport was often used as an alternative to the American airbase in the north. After the war, foreign tourists and an expatriate population continue to support the industry. The concentration of prostitutes serving a western clientele, particularly in the area known as Fuente Osmeña, has led to a focus of AIDS education efforts in this area, a marginal part of the city that is viewed as requiring medical attention.

In Manila, by the same token, the mayor chose to close down its famous sex tourism district, which forced the sex industry to go underground or to move to municipal districts in other areas. This forced a few establishments into distinctly marginalised positions, remaining open only furtively and subject to frequent raiding (Law, 2000:38).

---

group who form a minority in all European countries, the Middle East, Asia and through the Gypsy diaspora into the Americas, Australia and New Zealand. In many parts of the world, they have been racialised, discriminated against and excluded from mainstream society. They are culturally constructed as distinctive in their nomadic habits and in their attitudes to work, characterised by an avoidance of wage labour, a valuing of opportunism and flexibility (including the opportunity to exploit the dominant economy) and an integration of work with residence and family. These cultural attitudes and habits, and the accompanying negative constructions of them as lazy and shiftless, have made it difficult for them to be absorbed into sedentary societies where home and work have become distinctly separate spheres. Traditionally, they have been constructed as both romantic and deviant, as simultaneously objects of desire and derision. They are stereotyped as lazy, devious and dirty. Government agencies, the media and other decision-making elite have attempted to draw boundaries in which Gypsies are impure, deviant and transgressive and the rest of society is deemed to be pure and virtuous.

In Europe, there is a long history of attempts to evict, control and separate Gypsies from the rest of society (Sibley, 1999:147). The Nazi regime in Germany attempted genocide, while the Czech regime's denial of citizenship rights in 1997–98 has caused considerable emigration resulting in numerous claims for refugee status. Sibley (1999:147) comments: 'In modern industrial-ised nations the more general objective has been to settle and contain Gypsies, to remove them from locations where they are perceived as a non-conforming outsider group, violating space valued by the settled society, particularly resid-ential space.' Boundaries have been constructed in an attempt to purify space of this deviant and potentially transgressive group (Sibley, 1988).

In the English city of Hull, Gypsies have been present in roadside camps and seasonal movements for at least 150 years. In an attempt to control the deviant group and the reactions of local ratepayers, the City of Hull allocated two marginal spaces for Gypsy camping grounds. One was located in a re-sidual polluted industrial site previously declared unfit for residential use; the other in an old quarry site subsequently used for rubbish dumping. In both cases, a minority defined as morally polluting was allocated a space that was physically polluted. The sites were both physically isolated from the city and strongly controlled with caravans in regular rows without workspaces. In at least one site, however, the boundaries were not strictly controlled and the caravan dwellers were able to subvert some of the assimilationist and control-ling tendencies of the council, by introducing traditional work into the desig-nated residential environment (Sibley, 1999:149).

One group of Australian Aborigines who have been stigmatised more than any other Aboriginal group are the fringe dwellers – those living in self-constructed shacks on the edges of towns throughout outback Australia. In Australia, the term fringe camp refers to an encampment of people (usually Aborigines, but often including a few white people) living, at best, in under-serviced housing co-operatives and, at worst, in rudimentary shacks constructed of corrugated iron, wood and occasionally canvas, on the edge of a town (Gale, 1972). These camps often lack basic services, such as running water, electricity and conventional sewerage systems. Most camp residents spend the majority of their lives in the fringe camps. However, many inhabitants often leave the camps to find work – in response to the demand for labour on nearby cattle stations. Collmann (1988:6) claims that the movement of resid-ents in and out of the fringe camp incorrectly gives the impression that the camps are impermanent and their residents shiftless. In addition, he contends that the general impression of social decay in the fringe camps is frequently magnified by what outsiders consider to be abnormally high levels of alcohol consumption, violence, disease and other signs of disorganisation. According to Collmann (1988:22) fringe camp residents (or fringe dwellers) were often described as 'people who had lost their tribal ways and not yet completely learned white norms'. He argues that fringe dwellers in particular 'have been historically classed as deviants and have been the objects of overt techniques of therapy, social engineering and punishment' (Collmann, 1988:8). The location and landscape of these camps, literally on the urban margins, and often adjacent to noxious facilities such as rubbish dumps, reinforces negative constructions of Aboriginality (Cowlishaw, 1988).

Fear of the nomadic, the transgressive and the deviant is very common and the hegemonic reaction of institutions is to contain, control and assimilate. The example of hippies in England illustrate this tendency to label, vilify and control. In the hot summer of 1975, a hippy convoy roamed the back lanes and country roads of England. As the convoy grew, a moral panic ensued about the contravention of by-laws in relation to camping and flush toilets and of regulations about children's schooling. The convoy's participants were labelled and constructed in the media as hippies, wasters and deviants and they were eventually dispersed.

It is never clear how far temporary transgressions change power structures. Alternative lifestyles and landscapes, such as hippy communes in the backblocks, exist in many remoter areas. The town of Nimbin in northern New South Wales is a place where hippy lifestyles flourish, where marijuana is smoked openly and where income is derived from some cottage industry, welfare payments and spending by tourists. The relatively isolated town exists as a backwater; the surrounding hills and forests contain numerous communes and religious communities living a relatively sheltered and self-sufficient existence but having only desultory commercial contact and occasional conflict with various forms of authority. It is only rarely that some item of news brings the marginal to centre-stage for a brief moment in the spotlight.

While some marginal landscapes evolve or are established by the will of marginal peoples, others are marked out for them. Workhouses, or poorhouses as they used to be called, represent marginalised landscapes particularly established for the poor and the disabled and reflecting the strength of social constructions in effecting physical marginalisation and social isolation. In nineteenth-century Britain, institutions were established to confine poor and disabled people (see Box 6.3). Workhouses gathered the 'objects of charity', including the insane, the sick, the handicapped, the unemployed, into 'a single enclosure' (Durkheim, 1964:188), a 'pen of inutility' (Higgins, 1982:202), a 'country of confinement' (Foucault, 1979). These institutions kept disabled people away from formal public spaces in the nineteenth-century industrial city and, in that manner, marginalised them (Ryan and Thomas, 1978; Davis, 1995). Most disabled people lived 'invisible lives' in private homes and public institutions (Oliver, 1990). Such disabled people in the Victorian city became subjects or, as Gleeson (1999:119) pointed out, 'objects' of 'socio-cultural abjection', cast off, degraded and excluded from the lives of the 'normal', the employed, the able-bodied. They were 'expelled' by oppressive powers of the able(d), the rich and the powerful because they threatened the socio-spatial boundaries of normality (Sibley, 1995:18). Ironically, in the case of the Victorian workhouse, many of the buildings constructed were outwardly palatial in appearance, to give the impression of universal affluence and normality.

Box 6.3   The workhouses of Victorian England

Charles Dickens, the famous author of Victorian England, wrote poignantly, cynically, ironically and, oftentimes, scathingly of workhouses constructed for the poor. In *The Adventures of Oliver Twist*, for example, he wrote, with a huge dose of irony:

> The members of this board were very sage, deep, philosophical men; and when they came to turn their attention to the workhouse, they found out at once, what ordinary folks would never have discovered – the poor people liked it! It was a regular place of public entertainment for the poorer classes; a tavern where there was nothing to pay; a public breakfast, dinner, tea, and supper all the year round; a brick and mortar elysium, where it was all play and no work. 'Oho!' said the board, looking very knowing; 'we are the fellows to set this to rights; we'll stop it all, in no time.' So, they established the rule, that all poor people should have the alternative (for they would compel nobody, not they), of being starved by a gradual process in the house, or by a quick one out of it. With this view, they contracted with the water-works to lay on an unlimited supply of water, and with a corn-factor to supply periodically small quantities of oatmeal, and issued three meals of thin gruel a day, with an onion twice a week, and half a roll on Sundays. They made a great many other wise and humane regulations, having reference to the ladies, which it is not necessary to repeat; kindly undertook to divorce poor married people, in consequence of the great expense of a suit in Doctors' Commons; and, instead of compelling a man to support his family, as they had theretofore done, took his family away from him, and made him a bachelor! There is no saying how may applicants for relief, under these last two heads, might have started up in all classes of society, if it had not been coupled with the workhouse; but the board were long-headed men, and had provided for this difficulty. The relief was inseparable from the workhouse and the gruel; and that frightened people (Dickens, 1949 edition:11).

Perhaps one of the less glorious evidences of cultural diffusion was the emergence of similar asylums and workhouses in colonies like Australia. The Victoria (later 'Melbourne') Benevolent Asylum was one such poorhouse which started operations in 1851, with colonial government support. It was an opportunity for the colony's upper classes to 'demonstrate their peerless wealth and charity' through 'erecting . . . monuments to its own benevolence'. This was concretised in its physical structure and location, for, like other asylums of the time, Melbourne Asylum was built in palatial style (outwardly) and located prominently (Gleeson, 1999:123). It contributed to the myth that the colony did not suffer from poverty and fed the powerful social construction of the state of Victoria, Australia, as an 'affluent and civilized fantasyland, where even paupers got to live in palatial homes' (Gleeson, 1999:123). The realities of the institutional conditions, however, bore no relation to the grandeur of the outward landscape.

Part of the evidence of forced and reluctant placement in these institutions was that many resisted admittance and their long-term institutionalisation and left voluntarily, even if it was to return to unknown fates (see also Driver, 1989:269). Some escaped the 'austere and loveless mode of "care"' and the 'places of loneliness and brutality' that they had often been forced into (Gleeson, 1999:126).

This section on the landscapes of the poor has used a number of examples drawn from cities of Asia, Australia, Europe and north America. In all cases, the landscapes of the poor co-exist with the landscapes of the powerful and hegemonic and in many cases are defined as in opposition to it. These landscapes occur in a variety of circumstances, from those literally on the urban margins to those co-existing in prime space. Those literally on the margins include the enormous suburban estates and the fringe camps of Australian outback towns. Others, despite their apparently prime location, are stigmatised through social constructions of places and groups, including the occupants of 'ghettoes' whether racially defined or service dependent. In other cases, landscapes of prime space are subverted by marginalised groups such as the homeless, who require the use of prime space but who rarely overturn its hegemonic nature. In these cases, the marginalised groups tend to be invisible in the landscape unless a critical situation occurs; the example of the workhouses built in the grand style suggests a deliberate and constructed invisibility for the marginalised. The landscapes of the poor, with the exception of the subversive use of prime space, have been longstanding rather than transitory, embedded in discourses which fuel social constructions of the marginalised as deviant, dirty, immoral or lazy. The final section of this chapter is concerned with landscapes of transition, transformation and temporary inversions of the norm.

6.4 Liminal landscapes

Liminal spaces are broadly defined by Shields (1991) as being intimately associated with personal moments and movements of transformation: liminality occurs during 'moments of discontinuity in the social fabric, in social space, and in history. These moments of "in between-ness", of a loss of social co-ordinates are generally associated with religious experience' (Shields, 1991:83). Shields also says that, classically, 'liminality occurs when people are in transition from one culturally defined stage in the life-cycle to another' (Shields, 1991:83). Unlike the other marginal spaces considered in this chapter, which may be residual or obsolete or socially stigmatised, these spaces may be physically and socially attractive and are associated with personal transformation rather than exploitation. For example, places that Shields considers as liminal include Niagara Falls in the USA and the tourist resort of Brighton in the UK. Both places are environmentally attractive and both are associated with significant personal rites of passage such as honeymoons (passage from single to married) and even suicide (passage from life to death). They may also be associated with temporary inversions where people adopt different personas and normal social hierarchies are inverted.

The notion of liminality, as defined by Turner (1974a:156; 1974b), is borrowed from Arnold van Gennep's (1909, reprinted in 1960) *rites de passage* or rites of passage which accompany a change of state or social position. The concept of liminality has been used to understand religious pilgrimages: Turner (1974b:166–230) refers to pilgrimages as liminal phenomena; by extension, pilgrimage landscapes are liminal landscapes. As he illustrates, using a range of examples from pilgrim celebrations of the birthday of the deified General Ma Yuan in first century AD (Yang, 1961:89) to Malcolm X's (1966) pilgrimage to Mecca, pilgrimage is a liminal time during which 'all the pilgrims are on the same level' (Turner, 1974b:171). To quote Malcolm X:

> During the past eleven days here in the Muslim world, I have eaten from the same plate, drunk from the same glass, and slept in the same bed (or on the same rug) – while praying *to the same God* – with fellow Muslims . . . We were *truly* all the same (brothers) – because [of] their belief in one God . . . (quoted in Turner, 1974b:168).

Similarly, the concept of carnival, as a temporary inversion of social norms, is related and has been more fully discussed in Chapter 5. Hence landscapes of carnival may also be considered liminal landscapes. Using an extreme interpretation of the liminal, deathscapes may also be considered as liminal places, the place of repose of the ultimate personal transformation. They are imbued with special meanings and a certain amount of fear (see Box 6.4).

Box 6.4 Landscapes of the dead, cemeteries and cruces

Landscapes of the dead are both marginal and contested. In western societies, the dead are buried in large cemeteries located on the margins of cities. As the cities grow, cemeteries become surrounded by urban development. Instead of being locationally marginal, they become marginalised through their land use, places to be avoided except for the necessary burial and remembrance of the dead and maintenance of the grounds. Death is associated with fear and grief because of its link with the non-material world, unknown quality and apparent finality. Graveyards have become known in the popular imagination as 'spooky' places inhabited by ghosts and frequented by grave robbers. It is clear that in death as in life, landscapes are hierarchically ordered and socially segregated. In the ghost mining town of Kuridala which flourished briefly from c.1900–1922 in remote western Queensland, the only gravestones commemorate the mining managers and landed classes, men killed in mining accidents (whose gravestones were paid for by the mining company) and the occasional stone recording an infant death. Other graves are marked by mounds or small fences or wooden crosses, while many more must lie unmarked.

In recent years, more individual indicators of death have become frequent markers in the landscape. Hartig and Dunn (1998) describe the proliferation of individual memorials along the roadsides of New South Wales, a pattern also evident elsewhere, such as in New Zealand and in Mexico (Henzel, 1989). Each of these memorials commemorates a roadside fatality, usually of a young

(continued)

(continued)

male driver. They usually take the form of a simple white cross with a wreath or flowers and occasionally some other personal items. These crosses have evoked a variety of official responses. In New South Wales, the policy is now to allow them to remain, both in recognition of their personal memorial and also as a salutary reminder to other drivers. They are, however, landscapes imbued with multiple meanings and are surrounded by 'contradictory discourses condemning and condoning youth machismo' (Hartig and Dunn, 1998:5).

In countries such as Singapore and Hong Kong where land space is extremely limited there has been a pronounced move away from the burial of the dead to a greater use of cremation and second burial as space-saving measures (Teather, 1998:22). This has required a significant cultural shift in values and behaviour. Traditional Chinese cemeteries reflect a particular cosmological view of the universe and carry complex symbolic meanings (Teather, 1998:22–3; Yeoh, 1999:242). Traditional cemeteries were built to have auspicious *feng shui* but columbaria containing ashes more reflect the views of efficient and utilitarian bureaucracies. In this example, space and landscapes are clearly contested issues.

Deathscapes are not only space-using phenomena but are redolent with symbolism, whether it is of the meaning of masculinity or the construction of nationhood (Jeans, 1988). As such they arouse strong emotions and generate their own myths. Deathscapes are a useful vehicle to study 'the social constructedness of race, class, gender and nature; the ideological underpinnings of landscapes, the contestation of space, the centrality of place and the multiplicity of meanings' (Kong, 1999:1).

As intimated earlier, one of the most famous liminal spaces is the Niagara Falls. Shields (1991) viewed it as a liminal space to which pilgrimages are made and where rites of passage occur. The public perception of Niagara Falls has changed over the last two centuries as accessibility has improved and the natural environment has become more controlled. In the early nineteenth century, descriptions of Niagara Falls were as a place of awe and majesty, a significant natural wonder, with unimaginable quantities of water crashing over a horseshoe-shaped precipice, both terrifying and remote. As general accessibility to the area improved, it became a destination for honeymooners and also a significant location for suicides (the ultimate rite of passage) and for curiosity events, such as tightrope walkers performing the amazing feat of balancing all the way across the falls. In this place, the natural beauty and wonder makes it a place of desire and pilgrimage, where boundaries are crossed and identities and states can easily be in transition. Although the construction and normative view of the falls has changed and is now a site of mass tourism, it is a place of such unusual power and beauty that it provides an environment in which people can shake off their place-bound past identity and re-emerge in a new state, baptised, initiated, invigorated.

Similarly, the English seaside town of Brighton has been analysed as a liminal space which has changed its identity from the time of its development as a resort town in the Regency Era. Shields (1991:73) argues that: 'As a place Brighton came to be associated with pleasure, with the liminal, and with the carnivalesque.' In the 1780s, the Prince Regent was brought to the fishing village of Brighton to bathe in the sea water, for health purposes. As a consequence of royal patronage, the town grew extraordinarily rapidly as a beach resort. The pleasures of the beach were heavily disguised and legitimated by the medicalisation of the sea-bathing practice and its organisation through a complex set of rules. The beach was viewed as a physical threshold, a limen, outside the control of civilisation, with a possibility for appropriation to become socially marginal. Brighton beach provided the setting for a possibly miraculous and life-changing transition and as such became extraordinarily popular as a social centre and trendsetter in fashion.

The aristocratic Brighton season was followed by a period of mass patronage after the 1840s when train services enabled day trippers to take seaside excursions. Shields, refreshingly, sees this not as a ritual escape from factory work to keep the masses quiet, but as an 'enactment of alternative, utopian social arrangements' (Shields, 1991:91). The mass movement inevitably brought with it associations not just of pleasure, but of illicit pleasure and vulgarity, codified in the 'seaside postcard' where 'taking the air' becomes a form of innuendo and moral figures such as policemen or middle-aged women are scandalised by risqué behaviour. Brighton became synonymous with the 'dirty weekend' and a location of marital transgressions. In later years, Brighton acquired a more seedy image with criminal overtones; by the 1960s, beach fights between mods and rockers exemplified the liminal breakdown of social order. A construction of danger was added to the other place images of the city as decadent, risqué and glamorous, reinforcing Brighton's liminal place myth. Shields (1991:112) concludes that Brighton is located in an imaginary geography of Britain 'as a liminal destination, a social as well as a geographical margin, a "place apart"'.

A similar framework has been utilised in an analysis of the Australian Gold Coast, which annually hosts 'Schoolies Week', a rite of passage for young people who have just left school and will be going on to work or university the following year (Winchester et al., 1999). The Gold Coast is a liminal space, the easternmost point of Australia, at the borders of land and sea, New South Wales and Queensland. It is a product of tourism urbanisation (Mullins, 1993), a city that is marketed on sun, surf and sex. It was a popular destination for honeymooning couples in the post-war years and acquired a reputation, albeit less famous than Niagara Falls, as a place to go for this particular rite of passage. Schoolies Week is a major pilgrimage for thousands of school leavers each year, celebrating the end of school and engaging in bodily excesses of alcohol and sex, as a form of ritual carnivalesque release. For the Gold Coast, Schoolies Week is a major income earner, but is also a turnoff for the many families who form the biggest group of non-returning visitors.

The schoolies knew that their trip to the Gold Coast was not only a physical separation from home and school, but also a symbolic separation from school to the world of work. One boy, Adam, commented:

> It's sorta like a break between, like next week I'm gonna be starting working, so it's sort of a step from out of school into the workforce sorta thing. It's like a break in between sorta thing (Winchester *et al.*, 1999:65).

Others talked about their week as 'freedom', 'no nagging', 'a party, to get away from our parents'. During their time at the Gold Coast, they enjoyed a period as liminars, between states. Schoolies Week was for the participants a time when normative practices and performance codes were in abeyance. The freedom of anonymity was important for students who felt that away from home 'we're allowed to trash the place' (Winchester *et al.*, 1999:67). The liminal behaviour has numerous anti-social elements such as 'discharging fire hoses and extinguishers, broken bottles thrown in swimming pools . . . vomiting over top floor balconies, writing obscenities on walls and trashing rooms' (Winchester *et al.*, 1999:67). It also had numerous bodily elements where the school students pushed their bodies to new experiences and limits, particularly in the consumption of alcohol and the pursuit of sex. Two girls commented on the combination of bodily experiences:

> *Anne*:   Cause I've got a boyfriend at home, I'm not supposed to be picking up while I'm here, but I thought, you know, you're gonna be drunk . . .
> *Friend*: If it happens, it happens.
>
> (Winchester *et al.*, 1999:69).

Nonetheless, the students themselves imposed order on their disorderly days and nights, in what could be considered as 'a celebratory inversion of their school lives: out at night, sleep late, go to the beach, wander and shop on Cavill Mall, perhaps visit a theme park one day, go out at night to clubs or to "spectate" on the crowds on Cavill Avenue' (Winchester *et al.*, 1999:66). Cavill Mall and Cavill Avenue become liminal landscapes where the rites of passage are most publicly on display. The Gold Coast event coordinator for Schoolies Week commented that Cavill Mall was a 'seething mass of humanity, and it's basically fifty per cent of the population watching the other fifty per cent' (Winchester *et al.*, 1999:72). The spatial order of the dominant culture is suspended during this period of transition.

In 1996 the Gold Coast City Council attempted to take control of what they saw as an increasingly dysfunctional event that was giving the area a bad name. Strategies they employed included an event coordinator to organise 'safety-valve' events and performances and a public relations company to manage publicity (Winchester *et al.*, 1999:73). The repackaging of the week as a 'Schoolies Festival' under the theme 'Celebration and Control' has diminished the transgressive quality of the event for many schoolies. There is some evidence that the newly controlled Schoolies Week is losing popularity as school leavers seek other less controlled locations for their rite of passage.

A further type of personal transformation may occur within the marginal gay and lesbian spaces of cities. In Chapter 4 we outlined how heteronormativity precluded non-heterosexual performance in most spaces, often working through the experience or threat of homophobic violence. This generated a marginalisation of gays and lesbians (Valentine, 1993a:397–409). In response, gays and lesbians have developed their own, often marginal, spaces. In Chapter 5 we outlined some of the more well-known gay precincts that have developed in western cities such as San Francisco and Sydney. But most gays and lesbians do not live in these so-called liberated zones (Valentine, 1993b:246).

Linda Peake (1993:425) observed that most geographies of sexuality 'concentrate almost extensively on gay males' territories'. What little comment there had been on lesbians had been fairly dismissive of the importance of lesbian spatial dynamics or had rendered them spatially invisible (Bell, 1991:323; Peake, 1993:425). This dismissive treatment of lesbians can be traced to Castells' (1983:140) comment that in San Francisco:

> We can hardly speak of lesbian territory . . . there is little influence by lesbians on the space of the city. . . . Lesbians, unlike gay men, tend not to concentrate in a given territory, but establish social and interpersonal networks (Castells, 1983:140).

In a study of marginal groups in Paris, Winchester and White (1988) found a distinct but predominantly invisible lesbian geography, in part overlapping with the geography of male gay areas, but also in distinctive marginal quarters of the city. Subsequently, Sy Adler and Johanna Brenner (1992) 'found' that there were lesbian concentrations in American cities, but that they were different from the male 'gay ghettos', such as the Castro district of San Francisco (see Chapter 5). A series of studies has subsequently refuted the suggestion, in many ways sponsored by Castells, that lesbians lack space and community (Adler and Brenner, 1992; Peake, 1993; Rothenberg, 1995; Valentine, 1995). These authors announced that they had found a lesbian landscape and community:

> It seems clear that there is a spatial concentration of lesbians in our city, a neighbourhood that many people know about and move into to be with other lesbians. But the neighbourhood has a quasi-underground character; it . . . does not have its own public subculture and territory (Adler and Brenner, 1992:31).

The threat of violence has been seen as a major reason for the apparent absence of lesbian precincts (Adler and Brenner, 1992:32; Namaste, 1996:226–7; Peake, 1993:426). Valentine (1993a:409) speculated that the 'majority of lesbians interviewed modify their behaviour in order to conceal their sexual identity and so avoid antigay abuse'. She also described the lesbian 'scene' in an English industrial city as 'peripatetic' (Valentine, 1995:100). This lesbian 'scene' lacked permanency, with commercial and non-commercial landscapes temporarily transformed into lesbian spaces on specific days and at particular times.

The strategic concealment of sexual identity for lesbians is both spatially and temporally specific:

At different times and different places, such as at work or the sports centre, lesbians and gay men are constantly forced to decide if, how and when to disclose their sexuality. . . . lesbians negotiate multiple sexual identities over space and time (Valentine, 1993b:237).

Peake (1993:426) found that a lesbian residential area, identified in Grand Rapids, Michigan, was a landscape and a milieu in which 'lesbian friendships and networks can function' and, importantly therefore, it provided a 'location for the nurturing of lesbian identities' (Peake, 1993:427). Valentine (1993b:246) referred to the sense of community and collectivity which was produced within, and across, lesbian networks and how that often resulted in mutual resistance to oppression and also subverting of dominant heterosexuality. In these ways, lesbian spaces and venues may be categorised as liminal spaces, wherein women can take the 'first steps' towards coming out (Valentine, 1993c:110–12).

6.5 Summary

This chapter has reviewed a number of landscapes on the margin, ranging from the obsolete landscapes at a sub-regional scale to the landscape of personal transformation and rites of passage. In all cases, these landscapes are those defined as other to some aspect of power. In the first section, landscapes rendered obsolete by changes in capital (and some of their subsequent reinventions) were considered, recognising that changes in physical landscape require and feed into changes in discourse and social construction. The bulk of the chapter considered landscapes occupied by groups constructed as marginal, the poor, the racially identified, the disabled and the unemployed. In many cases, however, the landscapes here designated as marginal are integral to the capitalist enterprise and the pursuit of hegemony. Many have been stigmatised for long periods, while others are more transitory, invisible or subversive in nature.

Many liminal spaces are temporary spaces for personal transformation. Landscapes become the sites of liminality because their inherent qualities may be imbued with socially constructed notions of transformation and rites of passage. This personal transformation may contain elements of pilgrimage, which involve the movement of the individual away from the everyday environment or carnival which involves people and places in a temporary inversion of the moral order. Many such transformations involve rites of passage such as from youth to maturity or a finding of sexual identity. In these transformations, the body is always implicated. The landscapes of the body will be considered in more detail in the next chapter.

Chapter 7

Landscapes of the body

7.1 Bodies as social constructions

This chapter focuses on the landscape of the body, a scale of study relatively unfamiliar and new to geography. It is primarily because of the recency of geographical enquiry at this scale that this 'landscape' is separated from other social constructions. Haraway (1990:222) argues that bodies, and by extension their clothes, hair and adornments, are 'maps of power and identity'. Bodies can reflect power; bodies are also effectively used as resistance. The social inscription of bodies helps categorisation and identity, whether as powerful, mundane or marginal. Some of this chapter takes geographers beyond their normal comfort zone to consider bodily landscapes of people who are physically disabled or whose bodies are tattooed, marked or concealed or whose bodies mark marginal and dissident identities.

The naturalised assumptions of the categories of bodies is easily demonstrated when we see someone familiar in a different disguise: our fellow students dressed to the nines on the night of their graduation ball, the corporate businessman who sheds the business suit for joggers and sloppy tracksuit, the friend who suddenly appears with bleached hair after 29 years of natural dark brown. In each case, the body makes a statement about power and identity. Any change from the expected bodily appearance unsettles identities and the social relations surrounding those embodied identities are shaken.

It is only recently that geographers have begun to study the body seriously. Most cultural geography has been undertaken at local and regional scales, rather than at the individual scale of 'the geography closest in' (Longhurst, 1994). The academic interest in the body derives largely from the work of feminist scholars such as Elizabeth Grosz who argue very strongly that there is no such thing as an essential femininity but that women's identity is constructed through and on their bodies (Grosz, 1987:1–2). The body is socially constituted rather than biologically determined (see Chapter 2 on social constructivism and environmental determinism).

Changing control of the body demonstrates its highly political nature. For example, in Afghanistan, before the Taliban takeover of the mid-1990s, many urban Afghan women worked in schools, hospitals and offices, wore brightly coloured clothing and were actively involved in public life. The Taliban, overthrown in late 2001, controlled, restricted and contained female bodies:

women were prohibited from working and from moving outside their homes in public, girls were not allowed to go to school and women could be summarily punished for exposing any part of their bodies, limbs, faces or hair and even for wearing white socks or heeled shoes which are considered attractive. Control over the female body has been used to construct a particular identity for women, which is socially constituted and highly political. This circumscription of women in public life is not an essential attribute of women, but is one that is imposed by others through control of their bodies.

Clearly, it is almost impossible to talk of 'the body', as bodies come in various genders and are coloured and shaped in different ways. Nonetheless, western philosophy has tended to separate 'the body' from 'the mind'. This dualism runs in parallel with other dualisms, such as nature/culture, passion/reason, secular/sacred, object/subject and private/public. These divisions have been conceptually and historically sexualised and valued (Gatens, 1991), with women and femininity accorded a negative value in association with body, nature, passion, the secular, object and private, while men and masculinity have been accorded a positive value in association with mind, culture, and reason, sacredness, subject and public. The association of men with reasoning and science has positioned women as irrational and in need of control; control over women has often occurred through controlling women's bodies. Feminists have resisted this through academic discourse, socio-political practice and individual assertions of control over their own bodies. An example of this might be a western woman with cropped hair, wearing overalls and Doc Martens, who is making a purposeful individual statement challenging accepted notions of feminised bodies. In Afghanistan, during the period of attempted total control of women's bodies, a film made under cover in mid-2001 by Saria Shah for the BBC showed women giving secret lessons to girls and private beauty parlours where they made up each other's faces, resisting in small ways under threat of flogging and imprisonment.

The body is not a fixed essence, but is located in a network of political, socio-economic and geographical relations. As bodies are not fixed, their forms and functions can be redefined and contested and the body's place in culture can be constantly re-evaluated and transformed. The Afghan re-evaluation of women's bodies just outlined has occurred as a result of hegemonic processes rigidly imposed by a misogynist patriarchal government. In other circumstances, such a redefinition can be undertaken by individuals, through choices such as body building, slimming, body art and fashion. Individual bodies, as geographical places, are set in a network of local and wider social relations. An individual choice to lose weight, such as a widely publicised slimming campaign by President Clinton of the USA, was influenced by his own health considerations, but also by adverse media comment and by a societal valuing of slimness and fitness (as well as whiteness and maleness).

Bodies may therefore be viewed by geographers in a number of ways. Bodies are landscapes constructed within a network of social relations. They are therefore surrounded by assumptions contained in a variety of discourses. Bodies are also sites of resistance, which may occur collectively or individually.

According to Grosz (1987) and Longhurst (1994), bodies are 'the site of social inscription'. Resistance is inscribed, written, marked on our bodies. Dorn and Laws (1994) suggested that bodies should not be considered as objects but that geographers should engage in a new body politics of access and mobility. In the rest of this chapter, bodies are considered as socially constructed; as landscapes of resistance and inscription; and as players in bodily performances and body politics.

Earlier chapters of this book have argued that cultures and landscapes are socially constructed and that as such they are enmeshed in ideologies which shape them. Furthermore, they are not givens or essences, but are in constant phases of becoming; identities are fluid and continually negotiated. This evolution of identity is even clearer for the landscape of the body, through the organic processes of growth, development and ageing. In considering bodies as socially constructed, this chapter focuses on the constructions, the discourses through which they are reproduced and ways in which those constructions are played out or resisted. In considering bodies as socially inscribed, the focus of study is the bodies themselves, their corporeality and the lived experiences contained within them.

7.1.1 Bodies of power and perfection

The dualisms outlined earlier indicate clearly that bodies deemed to be powerful are socially constructed to reflect other naturalised hierarchies, of class, race, age, gender and religion. A legacy of imperial and colonial history is the power imbued in the bodies of older rich white Christian males. Bodies of power have always been identified by their clothing and adornment; it is easy to bring to mind the robes of priests, the crowns of kings and princes or the wigs and gowns of judges and magistrates. In modern capitalist patriarchal society, the robes of power are the business suit and, formerly, the bowler hat and umbrella. McDowell and Court (1994) in their article about merchant bankers showed a picture of senior corporate bankers, be-suited, be-spectacled, corpulent and balding, sub-titled 'the power of the father'. Interestingly, the younger bankers – 'the young turks' – also had bodies of power but which were generationally defined: slim, athletic, toned in the gym and more casually attired.

Both these stereotypes reflect the power of hegemonic masculinity (Connell, 1995). However, as we have intimated in earlier chapters, other types of masculinity are categorised as inferior. In particular, homosexual masculinities are seen as problematic and as subordinate to the heterosexual. Similarly, bodies which are racialised are also viewed as subordinate. The racialised body is one which has been powerfully and historically constructed through imperial geographies and a history of place and racial hierarchies. In such categorisations, the white (male) body is seen as the norm and bodies of other colours are constructed as different.

The perfect body of Magic Johnson, the American basketball mega-hero, emphasises how the body is the site of competing discourses of masculinity, race and sexuality. Magic Johnson had been boosted by the media to superstar status and his well-known smile had become 'a major tool for the advertising and promotion of Pepsi, Nestlé, and Target, a nationwide department store chain' (Rowe, 1994:11). His disclosure of his HIV positive status was constructed by the media as a sporting calamity and the diseased nature of his body was surrounded by complex discourses emphasising hegemonic racial and sexual hierarchies (Rowe, 1994).

As a black male, both his body, seen as a model of athleticism, and his status as sports hero, were compromised by institutionalised racism (Rowe, 1994:7). In simple terms, his status as a sporting hero with a model athlete's body would have been perfect but for his blackness. In terms of the dualisms outlined earlier in this chapter, the categories of white, male and heterosexual are hegemonically dominant. His status as a powerful male, albeit black, is reinforced by his sexual conquests and his denial of any homosexual tendencies, which have been categorised by Connell (1995) as a type of masculinity subordinate to the heterosexual. His lived experiences, where he claims to have sexually 'accommodated' 2500 women, were given meaning through these discourses. In particular, the media reporting reproduced a racist stereotype of black male sexuality (which is constructed as dangerously voracious, with overtones of animal instincts), a denial of the derided masculinity of homosexuality and a demeaning of women as mere objects of sexual gratification, in focusing on the construction and dismantling of the heroic status of Johnson's body.

Such differences are analogous to those identified by Jackson (1991:204) in his discussion and deconstruction of British Army advertising where images of two young men are counterposed against each other under the heading 'Is the Army looking for more recruits? Yes and No'. Jackson (1991:202) makes the following comments on their bodily appearance and clothing:

> Both men appear to be fit and healthy; both are casually dressed in the style of the times. Yet, visually and verbally the advert makes clear, the army is only interested in the man on the left: he is depicted standing upright, not slouching; in smart, casual attire (with incredibly well-pressed trousers), not in the slightly effeminate clothes of the more street-wise youth on the right. Even the details are revealing: one has carefully combed hair; the other is more tousled; one wears trainers (suggesting a healthy liking for outdoor sports), the other wears less robust open sandals.

These groups of young males are surrounded by discourses of fitness (and misogyny) as part of masculinity. However, in the case of the Army recruits (see also the 'clubbies' later), this fitness is grounded within other narratives of community service and control, order and discipline. In the case of the Army rejects (see also the 'surfies' later), individuality and resistance to conformity replace discipline and control. These discourses which contribute to the identity of the groups are socially inscribed on their bodies. Skin and clothing, hair length and style all embody self-perceptions of identity and constructions of social groups. Bodies are both the site of identity and the site of resistance.

7.1.2 Perfect(ible) bodies

The discussion thus far implies that there is an ideal body or bodies, against which other bodies are measured and found wanting. This ideal body is clearly a construction specific to particular societies. The desirable female body is an object constructed by male desire but also by capitalist exploitation. Women, in particular, have undergone 'self-torture, self-deprivation and self-mutilation . . . as they attempt to cope with hegemonic beauty ideals' (Gimlin, 1996). The beauty ideals are reproduced through a number of mechanisms including advertising: advertisements in slimming magazines aim to induce changes in women's spending and behaviour in order to achieve a new, better body:

> If an overweight woman eats Ryvita as a part of her calorie-controlled diet she'll actually look like the model in the bikini whose expression makes the biscuit seem delicious as well as slim inducing . . . The advert makes the woman dissatisfied with her own body so that she desires another (Day, 1990:51).

The desired female body has changed shape from the voluptuous Renaissance figure to the waif-like supermodels of the 2000s. However, such a preference is not shared by all women. A recent survey of young women in the USA showed huge contrasts between black and white views of the female body (Parker *et al.*, 1995). While 90 per cent of the white girls expressed dissatisfaction with their bodies, over 70 per cent of the African–American girls were satisfied with theirs. The white girls saw perfection as 110 pounds and more than half of them had dieted in the previous year, whereas the black girls valued inner beauty and style and more than half of them felt it was better to be slightly overweight than slightly underweight (Parker *et al.*, 1995:105–107). Clearly, the ideal female body is both socially constructed and culturally contingent.

The female body depicted most commonly in advertising material is 'tall, slim, young and white' (Winchester, 1992:149). Similarly, the ideal body shape for men is deemed to be 'young, lean, muscular'. In responses to a survey conducted by Mishkind *et al.* (1987:37–8), 95 per cent of college-age men expressed some dissatisfaction with their own bodies, while 70 per cent saw a discrepancy between their own bodies and their perceived ideal body type. Their research observed 'deeply entrenched cultural preferences towards mesomorphic males', preferences which had significant material consequences for the well-being of individuals (Mishkind *et al.*, 1987:39). There has been increasing recent emphasis on the overt mainstream portrayal of male bodies as sexual objects, using male bodies in the advertising process (Day, 1990:49; Jackson, 1991:205–206). Such uses may be overlaid with homoeroticism and exoticism of the black male body (Jackson, 1991).

The advertising of bodies reflects their commodification; bodies can be desired, bought, strived for, achieved. Day considers that 'the bodybuilder believes that his or her body is something to be developed and thus he or she conforms to an ideology of progress' (Day, 1990:53). The building of a hugely muscular body has changed from being the preserve of circus freaks and strongmen to being an end in itself, although one which he describes as purposeless (Day, 1990:54). The muscular body is specifically recreated not for power but for display as human strength is increasingly being replaced by technical developments. Day sees this purposeless reconstruction of bodies as an embodiment of the superficiality, distortion and purposelessness of consumer capitalism.

7.1.3 The social inscription of bodies – pregnant corporeality

Longhurst (1994) used a social construction approach in her study of pregnant bodies and their experience and use of shopping malls in Hamilton, New Zealand. Theoretical and practical implications arise from her study of pregnant bodies. On a theoretical level, the pregnant body itself upsets the dualistic thinking of our ways of knowing: one body contains another, one will become two and the boundaries between inside and outside, between self and other are challenged. This challenge to the established duality that the pregnant body presents has in part made it taboo. On a practical level, if women in general have somehow been equated with the 'natural', then pregnant women have been considered literally as hysterical (derived from the Greek *husterikos* 'of the womb'), beyond reason, prone to irrationality and therefore in need of surveillance and control (Longhurst, 1994).

Pregnant women have easily identifiable body shapes, especially in the later stages, which facilitate their identification and control. The medicalisation of pregnancy and childbirth has been a very obvious way of taking the control of women's bodies away from women themselves into the hands and machines of a predominantly male medical elite (Longhurst, 1994; Oakley, 1986). Women's pregnancies are hidden, pregnant bodies are denied access to public space, birth procedures have become heavily interventionist and women's individual identity has been submerged as 'the mother becomes merely the site of her proceedings' (Kristeva, 1980). As a consequence, pregnant bodies have been confined to the realm of the 'natural' and domestic and, as such, have found little place and voice in the discipline of geography. The place of pregnant bodies in Singapore society is the subject of Box 7.1.

Longhurst's (1994) study focused on pregnant women's experience of shopping in the Centreplace Mall, Hamilton, and on the discourses which constrain those experiences. The shopping mall, constructed in 1985, is a typical suburban three-level mall with approximately 90 shops, a cinema and various other facilities. As with many other shopping malls, the prime sites were occupied by women's fashion and clothing stores, often with women represented as sex objects, for example, in lingerie displays and perfume advertising (Longhurst, 1994; Winchester, 1992). The women interviewed by Longhurst (1994:219) reported that as their pregnancies progressed, they felt less and less welcome in such fashion stores:

> One respondent claimed that she received a 'frosty' response from the shop attendant when she entered a lingerie shop not to buy a feeding or maternity bra but to outfit herself in some of the latest 'sexy' underwear.

Box 7.1   The (self-)control of pregnant bodies in Singapore

Phua (1998) draws attention to the socially constructed nature of reproduction and pregnancy in Singapore society, arguing that there is clear evidence of power relations inscribed on the pregnant body. Pregnancy is not only a biological event but is infused with social meanings and ideologies, particularly those invested by the state, medical and cultural authorities (see also Chapter 3 on the ideology of motherhood and reproduction in Singapore song). At the same time, pregnant women engage in a range of small acts of resistance in everyday spaces as ways of thwarting attempts to determine what happens to their own bodies.

As an example, Phua (1998:219) illustrates how 'ideas about pregnancy and the pregnant body as put forth by medicine and its associated scientific know-how and routine practices are indeed socially constructed'. Western biomedical science subjects the pregnant body to intensive surveillance and regulation, from constant supervision to systems of measurement to laboratory tests, so that the body might be productive. Medical intervention and surveillance of the pregnant body is naturalised and normalised. Yet, 'by contrasting how pregnancy and the pregnant body are construed in western biomedicine and traditional Chinese medicine, the social construction of women's bodies become evident, and the supposedly "objective" medical knowledge of the pregnant body is brought into question' (Phua, 1998:220). Specifically, in Chinese biomedical discourse and practice, medical intervention in the form of various ante-natal care strategies are not emphasised, for the emphasis is on self-regulation in terms of balance and harmony, including 'harmonizing every aspect of regimen and conduct, encompassing the control of one's environment and the management of one's emotions' (Furth, 1987:13, quoted in Phua, 1998:126–7). As Phua (1998:127) argues: 'The salient absence of a well established ante-natal medical regime under this particular medical system (Chinese biomedicine) as opposed to the "normality" of ante-natal medical surveillance and regulation based on western models is indeed revealing in terms of how different medical systems approach pregnancy and the pregnant body.'

Even while Phua examines the exercise of power over and through women's pregnant bodies, she also argues that pregnant women sometimes struggle for control over their pregnant body, for example, through insisting on full knowledge of medical checks and procedures being performed on their bodies ('reflexive questioning') and through comparing medical treatment against bodily experiences and removing themselves from unsatisfactory medical attention.

In marked contrast to this, they reported feelings of welcome, initiation and affirmation in the new worlds of maternity wear and children's clothing and toys:

Their new body shape made them feel like welcome members of 'humanity' and that they were somehow approved of and accepted for fulfilling that role that women have always fulfilled, the carrying and bearing of children. They were met with affirmation. Their reproductive capacity was accorded social significance and value. They were similarly welcomed to the mall creche (1994:218).

The newly constructed mall environment, although presumably designed with women in mind as the main consumers in western society, in fact produced difficulties for women who were pregnant. Pregnant women needed to have more toilet facilities available with larger cubicles. The lack of such facilities actually inhibited women leaving the house. There were problems with stairs and escalators, which were either tiring or unsafe, and problems with inadequate and uncomfortable seats in the cinema and tavern. The cinema rows of seats were spaced too close together, while the tavern supplied only high backless bar stools which gave inadequate back support.

The discourses which surround these lived experiences are complex and intersecting. The built environment is a reflection of a social environment where pregnancy is hidden and private rather than public. Pregnancy has been constructed as a medicalised condition in which women are expected to behave in certain ways and to confine themselves to appropriate places and activities. Taverns are constructed as places for the consumption of alcohol (and often tobacco) and are seen as inimical to the health of the foetus. Many parts of this public space are deemed to be inappropriate places for pregnant women. This discourse of appropriate activities and spaces for pregnant women is manifest through physical difficulties in utilising the built environment, in the disapproval of others and in the social and physical discomfort of the women.

Other places are also deemed inappropriate for pregnant women. Longhurst recalls during one of her pregnancies, being 'dressed in waterproof gortex trousers and big boots . . . employed to operate heavy machinery' and as such was subject to surveillance and regulation by co-workers and 'loved ones' (Longhurst, 1994:221). Winchester's experience of running a ten-mile race when pregnant aroused interest and a certain amount of surprise, although prevailing attitudes to exercising pregnant bodies have become much more liberal and supportive in recent years. Contrariwise, social opprobrium of pregnant bodies smoking tobacco or using other drugs has increased. In complexly changing ways, pregnant bodies are subject to culturally defined constraints and naturalising discourses which are generally presented as protection of the unborn rather than recognition of the individuality of the mother.

7.1.4 The social inscription of bodies – surfers and surfies

Surfing bodies are icons of Australian beach culture. The overwhelming impression of surfers is of bodies which are muscular, tanned – and male. As such, surfing is seen as essentially a masculine activity, a taming of the vagaries of nature by the fittest and strongest with female participants ignored, marginalised or represented as sex objects (Fiske, 1989). Surfing is not, however,

Figure 7.1 The 'clubbies' in action: Australian surf lifesavers as team minded, community minded and disciplined; the bodies are male, muscular, uniformed, Speedo clad
Source: Telstra's *White Pages*, 1996.

a uni-dimensional activity, but is one which may be constructed in a number of ways. Figure 7.1 shows an Australian surf life saving club rescue team in action. This photograph was used as the cover photo for Telstra's *White Pages* 1996; the caption read 'The Redhead Surf Life Saving Club boat crew battles the surf and a true Australian tradition lives on'. The picture shows a disciplined boat crew: tanned, muscular, working together, uniformed, the epitome of coordinated teamwork. Pearson (1982:120) comments that the 'typical' lifesaver is thought to be 'conservative, community-minded, disciplined, fit, tidy, wear Speedos, have steady employment and drink beer at social occasions'. Such a stereotype accords with the lifesavers' role in enforcing the location of safe swimming 'between the flags', surveillance of the beach and surf from a watchtower, policing the beach and surf and effecting rescues. Surf lifesavers are part of a nationwide movement at club, branch and state levels, with a junior section known as 'nippers', increasing female participation and a complex and hierarchical series of competitions and events.

A contrasted construction and bodily representation of surfing is the group known as 'surfies' rather than 'surfers'. The commonality is the fitness, tanned skin and maleness of the bodies, but there the comparison ends. Pearson (1982:121) remarks that surfies are constructed as 'individualistic, hedonistic, undisciplined, untidy looking, long hair, unemployed and use drugs for kicks'. The growth of surfing as a part of Australian male youth culture in the 1960s occurred at a time of increasing mobility and affluence. Many surfers did not want to be confined to one beach and this hedonistic attitude resulted in increasing conflict between 'clubbies' and 'surfies'. The surfie, complete with combi van, wet suit, designer-label gear (Rip Curl, Hot Buttered, Billabong, Hang Ten, etc.) and all the accessories, adopts a surfing lifestyle and sub-culture where catching the best waves is the prime purpose in life. Surf lifesaving, however, is more likely to be a weekend activity in a relatively structured and conservative lifestyle.

This difference in lifestyle is inscribed on the bodies of the two groups of surfers. The lifesavers' smart blue Speedos, their identifying caps in club colours, their short neat hair, are all redolent of an ethos of discipline and control as well as fitness and masculinity. The surfies' long sun-bleached dreadlocked hair is probably the most potent indicator of their resistance to convention, but the impression of fluidity and movement is accentuated by tagged zips on wetsuits, leg ropes and wrist and ankle straps for adornment. The differences in bodily appearance indicate the dualisms between control and resistance, order and anarchy, centre and periphery, the group versus the individual. Nonetheless, both groups are hegemonically masculine and treat women as inferiors; until recently many surf clubs have excluded women from active roles, while the surfing lifestyle and magazines accord women a clearly subordinate status as sexual and decorative objects.

7.2 Bodies as sites of resistance

Bodies can be used as a site of resistance against social pressures and social conformity. People who feel they have virtually no control over any aspect of their lives may still be able to assert control over their own body. In the last resort, people can commit suicide or can take a course of action which marks their bodies as sites of resistance. Such markings include the piercing of ears, noses, lips, cheeks and other body parts; the cutting or laceration of skin and flesh; and the marking of text in dye on the skin. The tattoos of the prisoners of the gulag marked defiance on their skin. Such markings reflect a disaffiliation from the structured and comfortable world of home and work and stand in opposition to the smooth, tanned, 'natural' healthy body of consumer society (Chalmers, 1988:151). Chalmers argues that tattoos represent a fixing of dirt and deviance in bodies, forming an unwashable stain, the stigma of the urban savage and the dangerous classes of the slums (Chalmers, 1988:152).

Many types of resistance are evident through fashion in clothing and grooming. Younger generations make their own statements about resisting conformity, whether it is through the long hair and kaftans of the 'flower power' era, the denim and boots of skinhead rebels or the op-shop dressing of grunge. Bodies are used not only to make statements about conformity, but also in other instances as markers of morality. In 1970s' Singapore, there was a severe concern that the 'decadent west', manifest in hippy clothing, drugs and 'yellow culture', was overtaking the cultural traditions of the Chinese, Indian and indigenous Malays. The Minister for Home Affairs took an active interest in nightspots and effected legislation that insisted on surveillance or closure (see also Chapter 3). The accompanying social construction of moral geographies focused on the micro scale and particularly on bodies. A common refrain was that all the nightspots where potential substance abuse would take place were characterised by youthful customers who had long hair (or wigs) and wore coloured T-shirts, denim jackets with badges and signs, bell-bottoms, loose blouses and jockey caps (*The Straits Times*, 8 July 1972).

Other marginalised groups, such as gay men and lesbians, have made fashion statements which destabilise the categories of masculine and feminine appearance, including the use of body jewellery for men and cropped hair for women. In each case, resistance to conformity produces fashions which become the hallmarks of identifiable sub-cultures. Furthermore, many of those statements of resistance have become incorporated into mainstream fashion, as the hegemony continually modifies itself in incorporating the margins.

Taking control over the body is seen very clearly in limiting food intake. Food asceticism is not a new phenomenon. Bynum (1987) discusses fasting by women in the Middle Ages as a way of giving service to God. Fasting was one of only a few options open to women, who had control over their bodies but little control over property. The thinking of the time also constructed women alongside the bodily and the human rather than with the spiritual and the divine. So while men could be priests or could renounce their property, women were far more likely to take vows of chastity and to torture and starve themselves in a renunciation and mortification of the flesh (Bynum, 1987). The transition from holy anorexia to the status of disease occurred in the late nineteenth century with concerns over the control of public health (Brumberg, 1988).

Naomi Wolf in *The Beauty Myth* (1990) argues that anorexia and/or bulimia affect a million Americans every year, while a number of countries have 1–2 per cent of their female teenage population suffering from food disorders. Feminists blame the prevalence of these disorders not only on strong gender socialisation, but on the capitalist–patriarchal structures which emphasise the sexual attractiveness and potential profit to be derived from the slim female body. Anorexia and/or bulimia, when people starve themselves, binge eat and vomit food and become thinner and thinner until vital organ functions are impaired, is not, however, merely an enslavement of young women to an unattainable ideal. Anorexia can also be an individual's way of resisting family pressures (punishing parents, for example), a way of suppressing sexuality (menstruation ceases) and a way of defining individual identity. Similarly, Orbach (1982) argued that 'fat is a feminist issue' because, for individual women, being fat is a way of controlling one's own body and resisting the pervasive patriarchal view that defines women by their body shape. In complex ways, the body is the site of social pressures, resistance and identity. It is the site of resistance against others and against itself.

Bodies can be used as elements of resistance in indirect ways, through subverting the categories of naturalised identity. Mention has already been made of the destabilising impact of lesbian and gay dressing. The transgressive nature of these bodily encodings challenges stereotypes. Similar challenges to categories occur through specific groups such as 'Bikers for Christ' who adopt the leather gear and powerful motorbikes of the tough-guy bikers, but who collect Christmas presents for disadvantaged children. The conflation of the 'macho' biker with the 'wimpy' 'do-gooder' Christian unsettles both categories.

7.2.1 Marginal bodies

Bodies, like other landscapes, are intensely political. They are a site of struggle and resistance. Dorn and Laws (1994) comment that medical geographers have objectified and medicalised bodies, treating bodies as unproblematic objects rather than social constructions. Sick bodies in particular throw into focus the 'ideal' body, the body that conforms to social norms. They argue that historically, bodies that do not conform have been confined, hidden in nursing homes, gaols and asylums (see also Chapter 6). Non-conforming bodies (Kristeva, 1982) have been denied access and mobility to public goods and places, and have been defined as different, other and inadequate. The problems facing non-conforming bodies are not merely physical ones of access or social discrimination, but are highly pervasive cultural and political constructions. Indeed, Gleeson (1995:11) goes further in arguing that 'disability . . . should be understood as a socially imposed state of exclusion which arises from the specific material organisation of society'. In other words, he argues that disabled people are excluded from public transport, from gainful employment, from the material goods of society because of the way society organises its activities and space. Gleeson (1995:13) argues that the core experience of these lived bodies is material exclusion and social marginalisation and that 'attitudes, discourses and symbolic representations are the product of the social practices which society undertakes in order to meet its basic material needs'. In arguing for a new body politics of access and mobility, Dorn and Laws (1994) challenge the material and cultural marginalisation of impaired bodies.

The body politics of access and mobility occurs at individual and at group levels. At an individual level, geographer Reginald Golledge (1997:391) has outlined the ways in which he 'reassembled his life' 'after the sudden onset of severe vision impairment (legal blindness). Golledge discusses the practical problems arising from his loss of vision particularly those in returning to the classroom and academic research. He lists major difficulties which include: cracking the print barrier; preparing illustrative materials; accessing maps and graphics; developing lecturing techniques; defining sets of research problems that could be pursued even with a disability; and rejoining academia through the usual means of conference presentations, journal submissions and so forth. Golledge (1997:395) summarises how he felt about trying to cope with the huge flow of visual information with a visually impaired body:

> The feeling of imprisonment was intense. Previously I had read widely in the books and journals of many disciplines. Being suddenly cut off from these resources and joys was a substantial shock. It brought my teaching and research activities to a standstill. . . . The road ahead looked obstacle filled, lonely, and desperate.

The 'cracking of the print barrier' came predominantly through techno-logical means, including optical scanning of print, a speech synthesizer and a vision enhancer (e.g. for displaying equations on screen at up to 16 times normal screen display size). Nonetheless, because of the time taken to listen rather than read, Golledge is unable to browse and explore new material with the enjoyment of the past. Similarly, lecturing techniques were adapted with technological solutions such as key words on a Walkman tape player, but the problem of coping with class interaction required a sighted teaching assistant to spot 'puzzled expressions, curious glances, or nodding heads' (Golledge, 1997:404). We focus on the material ways in which Golledge (1997:393) dealt with disability, because this is the way in which he views disability. He summarises his view of disability thus:

> In general, therefore, there are two different ways of looking at disability: first as a physical limitation and second as a limitation of the power to perform social roles. The first relates more to a person, the second relates more to society's attitude towards and treatment of disability. In this latter point of view it is assumed that disability results from forms of social and political discrimination. It should be clear that I reject this view as being a partial, abstract, academic or intellectual view with little basis in the reality of living.

Even more strongly in the abstract, Golledge comments, 'exception is taken to the red herrings of political correctness and power politics as relevant schema for examining disability concepts' (1997:391). This view supports that outlined by Gleeson earlier, that his disability is an exclusion from the real material world of information and marginalisation from academic practice. This statement clearly outlines his view of disability as being grounded in the specific material organisation of society. In this, Golledge is also making a personal and political statement from a position of relative power as a respected and established member of an intellectual elite.

Other commentators take a different perspective on the construction of bodily disability. Butler and Bowlby (1997:428) argue that 'the experiences of disabled people in public space always involve interactions between their bodily characteristics and social discourses concerning disability and public behaviour'. In taped intensive interviews with visually impaired people, it was revealed that the people with disabilities were acutely aware of media images and the opinions of others. These interviewees were 'highly self-conscious and self-critical about their appearance and behaviour in public' (Butler and Bowlby, 1997:423) and many made considerable efforts to look and behave 'normally'. They were also highly aware of the negative stereotyping associated with disability, particularly the association of physical disability with a lack of intelligence. The wheelchair or the white cane are the visible signs on the bodily landscape which give rise to negative social constructions. One woman, visually impaired from birth, commented:

> You do get sympathy and you're treated as if you're mentally deficient and patronised. Places like hospitals for example, one isn't allowed to be intelligent (Butler and Bowlby, 1997:424).

Another older woman, visually impaired late in life, said:

> They tend to think that you're not all there somehow or other [laughs]. And they talk loud to you or, er, ignore you altogether (Butler and Bowlby, 1997:424).

The interviews also stressed the great variation contained within the category of 'disabled', both in bodily function and in attitudes to impairment. Many people with disabilities are not passive, but actively resist categorisations and hurtful reactions by others. One woman was accused of being a fraud, but came back defiantly:

> I said I'd bet; I'd have a bet on with him. I said I'd get a glass, have a glass of water there. And you know I'd take my eyes out and say look there's my eyes. Now you hold one in one hand and one in the other and see how much you can see with them (Butler and Bowlby, 1997:429).

Such examples refute the social constructions of people with disabilities as unintelligent, passive and compliant.

Some categories of disability are deemed to be more acceptable than others, in that they are perceived as being closer to normality. Such hierarchies of acceptance have some constancy over time, although new categories have been added, while some have changed their acceptability ranking. Indeed the hierarchy as a whole may shift, reflecting societal changes in levels of acceptance (Dear *et al.*, 1997:466). One of the most enduring features of such hierarchies is the non-acceptance of mental illness and disability. Old age has become more 'acceptable', which may reflect changes in the way in which old age has been conceptualised over the last 20 years. New groups who have entered the hierarchies are the homeless and HIV–AIDS sufferers (Dear *et al.*, 1997:467). The latter are intensely stigmatised:

> Because AIDS remains a fatal condition, is associated with already marginalized populations, and can be transmitted through bodily fluids, it has caused an intensely negative public reaction. . . . people with AIDS were ranked as significantly less desirable than people with any other disability (Dear *et al.*, 1997:468).

Dear *et al.* (1997:471) conclude that the existence of such hierarchies and the changes within them are useful evidence that the construction of disability is contingent on and varies with time, place and culture. The landscape of bodily tolerance changes over time and from place to place. Bodies which are marked as 'Other' and categorised as different may have that difference emphasised by the ways in which the built environment is constructed. For example, a wheelchair user may be made to feel out of place in a building without ramps. A social construction of disability is reified in the landscape and in the material organisation of space.

The socially constructed nature of disability is evident from a case study of the mobility of disabled groups on pilgrimages to the Roman Catholic shrine of Lourdes (Dahlberg, 1991). Pilgrimages to achieve a change of state have been briefly discussed in the previous chapter; in this example Lourdes as the site of pilgrimage may also in some ways be considered as a liminal space (see

Chapter 6). This example of pilgrimage is included here because of its primary focus on the healing of bodies. Lourdes is the major place of pilgrimage for Roman Catholics, particularly for those in search of healing in a miraculous transformation of sickness to health. Lourdes was a place where the Virgin Mary, mother of Jesus, appeared in visions to a 14-year-old girl in 1858. At the site of those visions, a healing spring arose which has been the source of numerous miraculous cures. Pilgrimages to the holy place of Lourdes are structured into the annual lives of English dioceses and this small city in the south of France receives over 4 million visitors per year (Dahlberg, 1991:34).

In Dahlberg's study of one pilgrimage from the Diocese of Liverpool in the late 1980s, 4000 pilgrims travelled from the north of England to the south of France, including a large number of handicapped children, who travelled in relatively small groups and were cared for by doctors, nurses, clergy and a veritable army of voluntary untrained auxiliaries. These sick people had their place at the margin of society inverted so that they became the centre of attention: 'The hospitals, schools, institutions, all the hiding places of the deformed, retarded, and handicapped had been emptied out and their inhabitants swept triumphantly through places normally indifferent to their existence' (Dahlberg, 1991:41). Instead of being hidden from the public gaze, the handicapped acquired a public identity, where strangers competed and indeed paid for the privilege of working for them. These helpers washed, fed and entertained the children and gave them as much opportunity as possible to do what they wanted. The mass pilgrimage became chaotic and carnival-like with a party atmosphere stimulated by coloured hats and wheelchair races. As with other carnivals, however, there was an element of the temporary inversion of normal social relations (see also Chapter 5). In this inversion, sickness rather than health was accorded a valued position, bringing holiness through suffering, holiness through caring for the sick and inspiring hope for a miraculous cure.

This inversion rests on a particular construction of the body, based on the relationship between body and soul, the human and the divine. Catholics believe that Jesus was God, the divine made flesh in a human body. In the sacrament of Communion, the bread and wine that is consumed is the actual Body and Blood of Christ, not a representation thereof (this doctrine is known as the Real Presence). The body, as recipient of the divine is therefore seen as a vessel not to be tampered with. However, mortification and suffering and self-denial through the body is a way of achieving greater holiness. Care for the suffering is therefore also an act of holiness. This construction of the body as essentially inviolable but as a potential medium for salvation is seen in other teachings of Catholicism, notably restrictions on abortion and contraception, the practices of penance, fasting and abstinence and the practice of celibacy by priests. Dahlberg contrasts the Catholic conception of the body with that of Protestants who do not believe in the Real Presence (Dahlberg, 1991:46–7). At one extreme, Protestants may adopt a position where the body, because it is the not-divine, may be improved in any way possible (the Seventh-day Adventist view). At the other extreme, if the body is viewed as both other than and inferior to the divine, then medical treatment is refused as the body must submit to divine will (the Christian Science view). These particular constructions of the body bring a whole range of consequences beyond the medical to significant issues of the rights, duties, movements and actions of both abled and disabled bodies in a complex social politics of access and mobility.

7.2.2 Transgressive bodies

The use of resistant body politics may go further than challenging hegemonic authority into the transgressive. Bell *et al.* (1994) discussed the adoption by some lesbians and gays of quintessentially feminine and masculine dress and appearances. The so-called 'gay skinhead' and the 'lipstick lesbian' challenged naturalised assumptions regarding sexuality; of what is deemed to be feminine, masculine, lesbian or gay. The lipstick lesbian in adopting the high heels and make-up of the hyper-feminine confuses for the onlooker 'the object of derision' with 'the object of desire' (Bell *et al.*, 1994:42). Skinheads, whose bodies are already inscribed with resistance to convention, are further subverted as a category when seen to exchange tender kisses. Furthermore, such destabilisations of categories also challenge naturalised notions of the landscape as normally heterosexual unless queered (see Chapter 5).

These corporeal forms have been a target of criticism within gay and lesbian communities. The 1994 article in *Gender Place and Culture* by Bell *et al.* noted that the choice of the skinhead dress code – which has racist, violent (and homophobic) associations – forms a problematic identification for gay men. The political efficacy of such categorical confusions has also been a topic of controversy. Bell *et al.* (1994) questioned the political potential of such individual acts of transgression. They remarked, for example, that the unsettling nature of the lipstick lesbian is minimal unless the onlooker knows of the parody being performed (Bell *et al.*, 1994:33, 43). Kirby (1995:93) extended upon their concern and argued that the non-normative performances of gender may not be noticed by enough (heterosexual) people to have a resistant effect.

Bell *et al.* (1994:38–43) focused on how lipstick lesbians were able to 'pass' within straight landscapes and society, thus gaining heterosexual privileges. Valentine (1995:101) referred to those lesbians who were closeted or in networks which were invisible (see Chapter 6) as 'passing' within their heterosexist surroundings. There is in these descriptions an implicit dissatisfaction with those who pass – whether in their bodily performance or landscapes – because they do not overtly and constantly confront naturalised heterosexuality. Indeed, the lipstick lesbian parody of embodied conventional femininity may unsettle, but it could also serve to reinforce patriarchal structures and stereotypes.

As part of the radical feminist project, and much of foundational lesbian politics, a return to 'natural' or 'original' women was sought, one without the trappings of femininity such as skirts, high heels and make-up. The result was an 'androgynous look', typified by 'unshaven legs, short cropped hair, Doc Martens and baggy clothing' (Bell *et al.*, 1994:41). These performances were part of a project to construct a transcendent lesbian identity and performance (Valentine, 1993c:112). A prescribed look was generated and lesbians who did not conform were ridiculed and felt alienated as a result. Fiona McGregor (1996:34) complained that in the Australian scene there had developed a common construction of all 'dykes as card-carrying activists with greater political obligations than the general population'. Transgressing this set of strictures about the performance of lesbianism brought about a 'headmistressy rapping over the knuckles for disobeying the rules of the Lesbian School' (McGregor, 1996:34). Valentine found the same concerns being voiced by some lesbians in Britain. A fixed lesbian subject had been identified and gender performance in the 'lesbian scene' was strictly policed (Valentine, 1993b:245–6; 1995:103–4).

Nonetheless, many lesbians rejected the strict standards of performance as constructed, instead adopting what were considered hyper-feminine looks. Bell *et al.* (1994:42; see also Valentine, 1993b:245) pointed to the rise of the 'lipstick lesbian' or the 'designer dyke' in the late 1980s as evidence that the androgynous type had been undermined. The geographical analyses of these alternate lesbian bodily performances (Bell *et al.*, 1994:42–3; Chouinard and Grant 1995:146–7) have been less than charitable (see also Walker's (1995:72) critique of this literature). In their assessment of the political worth of the 'lipstick lesbian' performance, Bell *et al.* (1994:43) decided that this 'queer sexuality' was largely a resistance to the construction of earlier lesbian feminism. They further concluded that it did 'nothing to undermine patriarchy' and only contributed to the 'disintegration of the lesbian feminist project'. These debates within lesbian communities reveal that the body may be a landscape of resistance against hegemonic authority (heteropatriarchy) but also to the rules and norms of minority groups.

Much of the recent geographical scholarship on sexuality has revealed how 'public' space is regulated with the use and threat of violence (Namaste, 1996; Valentine, 1993a:407–11; see also Chapter 6). Sexuality and gender are policed through the landscape, non-heterosexual sexualities are made invisible and only 'culturally sanctioned gender identities' are tolerated (Namaste, 1996:238). Gays, lesbians and other non-heterosexuals become reticent to disclose their non-conforming or dissident sexuality in such environments (Kirby and Hay, 1997; Valentine, 1993a:410; 1993c:241–2). If the landscape is constantly and repetitively constructed as heterosexual then it is important to challenge the 'straightness of our streets' (Bell *et al.*, 1994:32, 44). The queering or sexual diversification of space is therefore critical. Duggan (1994:11) explained how the performance of non-normative sexualities – 'non-conforming practices, expressions, and beliefs' – is a form of political dissent. For example, the lipstick lesbian performance can be a personal, but nonetheless political activity in which the body is the landscape through which naturalised ideas about sexuality are challenged. Such personal challenges may confront heterosexist assumptions as well as those corporeal assumptions that have been generated within gay and lesbian communities regarding appropriate forms of dress and appearance.

7.3 Summary

In this chapter, the body as landscape and social construction has been considered. In particular, it is recognised that bodies are not only 'natural' but that they are constructed, imbued with meaning and intensely political. A body is a landscape which can be redefined and contested and which lies at the intersection of many complex discourses. Bodies may be seen as socially constructed, reflecting powerful ideologies, but they are also sites of resistance. Bodies which do not conform to the hegemonic norm may be marginalised but may also be transgressive of those norms. As such, they are landscapes of power, of the margins, of the everyday and of resistance. The geography of body landscapes reflects at the most intimate scale the cultural geography of landscapes outlined in this book.

Chapter 8

The role of *Landscapes*

8.1 Approaching landscapes

Cultural landscapes take many forms and play many roles. They are expressions of culture; and they help to maintain and reinvent ways of life. They are also implicated in power relations, as expressions of power, as well as embodiments of resistances to power. These different roles have been explored via geographers' different approaches to analysing cultural landscapes. Most fundamentally, these differing approaches have involved a varying interpretation of culture. In Chapter 2, we outlined the different ways in which geographers have approached the analysis of cultural landscapes.

The first broad type of approach we discussed was that of the environmental determinists, such as Ellen Semple and Griffith Taylor. For these geographers, culture was the outcome of the environment. In Chapter 2, we looked at examples of how landforms were the bases of creation mythologies and at how local flora and fauna can influence economy and society. We have shown in this book how built and natural environments can have dramatic effects on the cultures that exist within them. Physical landscapes can significantly influence culture. Many of the case studies we have drawn on certainly demonstrate how landscapes are a **means of culture**. The landscape can concretise or materialise culture. Landscapes facilitate cultural expression, maintenance and change, but we are not of the view that landscapes *determine* culture. In this we depart from the environmental determinists.

The Berkeley School of cultural geography had a very different perspective on the relation between culture and landscape from the environmental determinists. Geographers like Carl Sauer and Amos Rapoport had an almost polemic approach to that of the environmental determinists. Indeed, these Berkeley School geographers were cultural determinists. Their assumptions were that culture was a largely static way of life into which individuals could be sorted and categorised. As a superorganic entity (see Chapter 2), culture was imprinted onto landscapes. Most fundamentally, the Berkeley School drew attention to the **expressive role** of the landscapes. Key geographers of the Berkeley School were able to generalise the impacts of cultural groups, for example, by detailing the morphology of 'the Muslim city' (Chapter 2). There have been many examples in this book of the expressive or reflective role of landscapes. These include the familiar and increasingly ubiquitous markers of global cuisine, such as the Golden Arches of the McDonald's Family Restaurant (Chapter 3). Their (omni)presence reflects cultural globalisation. Graffiti, particularly the prolificacy of tags and gang graffiti, are respectively representations of global cultural trends and local cultural presence. We discussed how the landscapes of gay precincts provided a visibility and presence within heterosexist cities (Chapter 2). Sikhs in Singapore used landscapes and debates over specific landscapes to assert their national presence (Chapter 5). Similarly, bodies can be seen as landscapes through which culture is expressed. In Chapter 7, we looked at the corporeal expression of Christian faith through fasting and of varying types of masculinity in the surfies and surfers. Bodies are the sites for the expression of dominant cultural practice, including narrow constructions of femininity as well as dominant notions of order and purity (Chapter 7). But many of these case studies also informed us that landscapes are not just clear pages onto which culture is written. Landscapes have a key role in the process of culture. We do not accept the superorganic perspective on culture that was apparent within much of the most noted Berkeley School research on cultural landscapes.

The 'new' cultural geography of the 1990s has stressed the dynamic nature of culture. Culture is a process and landscapes play an important role in the **process of culture**. We have shown how global cultural trends are adapted and transformed by local circumstance; our case studies included cuisine and music (Chapter 3). Again, these examples and others pointed to how landscapes were a means of culture. In Chapter 6, we discussed how lesbian spaces were important to the recreation of community and identity – fundamentally providing a space for 'coming out', if only for short periods before closeting again to avoid violence and other forms of homophobic scrutiny. We explained how liminal landscapes, such as at Niagara Falls (USA/Canada), Brighton (UK) and Surfers' Paradise (Australia), were spaces for cultural change, including transformation and transgression. The destinations and paths of pilgrims can similarly be landscapes of profound cultural change, including for example, conversions or a deepening of faith and belonging. Everyday landscapes, such as street signs are used in attempts to concretise cultural hegemony or to reorient national identity away from a repressive colonial, socialist, elitist or racist past (Chapters 4 and 5). 'New' cultural geographers have drawn attention to the politics of the cultural landscape. Indeed, the influence of cultural landscapes on power relations has been a major theme of the case studies in this book. We comment more on the politics of landscape in the following section.

Another significant contribution of 'new' cultural geography has been the reassertion of theory. While Berkeley School geographers were clearly informed by specific theories, it is generally accepted that they rarely made their theoretical positions explicit (Duncan, 1990:12–13; Jackson, 1989:3; Price and Lewis, 1993:9–11). In this text we have drawn on two significant strands of theory. First, we have utilised a social construction approach to understanding culture (Chapter 2). Cultures do not just apparate; neither are individuals simply born into discrete cultural containers, through which they live their lives.

Cultures are reproduced, created and lived by current groups and individuals, both present and historic. Our second important theoretical strand has been the Gramscian view of power relations and the role of landscapes in maintaining and challenging power relations (Chapter 4). The case studies in *Landscapes* demonstrate that power is multiplicitous. Rarely, if ever, do dominant groups or individuals retain power absolutely. Indeed, they come closest to complete dominance only when they oppress with the *consent* of those who have been constructed as subaltern. As we show in the next section, cultural landscapes are important to the exercise of power.

8.2 Politics of landscapes

In Chapter 6 we made reference to the work of Iris Young (1990) on the forms of oppression. Our focus in that chapter was specifically on marginalisation. But throughout this book we have also demonstrated the role of landscapes in other forms of oppression. These include cultural imperialism, exploitation, disempowerment and violence.

The oppression of cultural imperialism involves the social construction of norms, specifically the identification of Others who deviate from the norm or Self. Our case studies have demonstrated how landscapes assist with such constructions. Particular landscapes – such as poor suburbs, Chinatowns or 'fringe suburbs' – become stigmatised as zones of disadvantage, incivility and danger (Chapters 4, 5 and 6). Once the landscape has become stigmatised, all inhabitants come to wear the myths of that place. The construction of those landscapes is transposed onto the people therein. They are constructed as deviant, uncivilised or generally inferior and these constructions are used to justify other forms of oppression, such as marginalisation and disempowerment. In circumstances such as apartheid in South Africa (Chapter 4), cultural groups are sorted across the landscape. The physical separation of Gypsies and blacks also assisted the construction of a normal Self and deviant Other (Chapter 6). Elite and minority groups have landscapes to which they have become symbolically associated. The less powerful are excluded and confined to their landscapes. Less powerful groups are deemed to be out of place when outside their zones of confinement. Elite groups rally to repel the homeless, sick and disabled from their places of prestige (Chapter 5). Gated communities define who is in and who is out (Chapter 4). Muslims were constructed as out of place in Sydney suburbs, prostitutes became aberrant and displaced by Muslims in suburban Bradford (UK) and Filipino cooking was prohibited in the Chinese kitchens of Hong Kong (Chapters 3 and 5).

Fundamentally, cultural landscapes help reinforce the notion of cultural difference. This was quite stark in many of our examples: the murals and graffiti of Belfast, the apartness of Chinatowns, the public housing estates versus gated communities, the Jewish ghetto, the cultural division of the colonial city (Chapters 4, 5 and 6). Often, these divisions give materiality to notions of racial, religious and class hierarchy. In other circumstances, cultural imperialism works through the silencing of groups. Constructions of national and local identity are quite selective in the people and landscapes that are incorporated. Official declarations, national songs and other cultural products such as film, make reference to landscapes that will reflect the experience and vision of those with power (Chapter 3). Confinement and institutional neglect can conspire to write many people out of symbolic existence. This is often engineered through the spatial confinement of the poor, ethnic minorities and the disabled to landscapes of neglect (Chapters 4 and 6). This allows the powerful to revel in their own sophistication, benevolence and superiority. The heterosexed nature of space aids the assumption that heterosexuality is a norm and that homosexuality is deviant (Chapter 4). Indeed, the presence of gay precincts may simply aid the construction that all other landscapes are heterosexual (Hodge, 1995). The construction of ordinary landscape as heterosexual, middle class and male all aids the social construction of the dominant Self in western societies (Chapter 4).

The physical sorting of cultural groups into separate landscapes, often along hierarchical lines, aids oppressions such as exploitation, marginalisation, violence and disempowerment. Having rich and poor suburbs is important to the division of labour in a capitalist society, as shown in the case of Manila (Chapter 6). Having gendered spheres of activity, female and male landscapes, reinforces the stereotypes regarding 'a woman's place' (Chapter 4). The construction of the feminine as domestic and the according of the home as a woman's place was fundamental to the exploitation of women. The cultural construction of gendered landscapes reinforced the free transfer of women's energies to men. Workhouses were landscapes in which the profound exploitation of the poor, sick and homeless was sanctioned as laudable (Chapter 6). We have dedicated a good deal of attention in this book to discussing how cultural landscapes are involved in the marginalisation of people (see especially Chapter 6). Clearly, separation and apartness are critical to making people marginal. The making of cultural landscapes can reinforce marginality, defining people as useless, an embarrassment or a burden on society. As we have shown (Chapters 4 and 6), separation and exclusion operate to dispossess, remove opportunity and generate dependency. Violence is sanctioned or tolerated against the less powerful if they move beyond their own landscapes. Gays and lesbians endure the fear of homophobic attack if they are 'out' in everyday space, women must negotiate the fear of sexual attack in public spaces and the police and other agencies are sanctioned by the state to use violence to move on the criminalised poor, 'threatening' youth and some ethnic groups. The cultural construction of landscape is used to legitimise violence.

Ultimately, cultural landscapes have a role in the allocation of power and control. They disempower and they empower. For example, monuments and monumental buildings are intended to demonstrate the power and status of powerful individuals and groups. In Chapter 4, we discussed how national monuments in Indonesia and Australia are used in an attempt to legitimise authority of the state. Religious monuments are also used to capture legitimacy for aristocrats. The attire of bankers and traders were referred to in Chapter 7 as the robes of power, another landscape of power. Hallmark events and the design of new towns have been revealed as having an overt focus on legitimating the decision-making elite (Chapter 4). Similarly, the re-identification of industrial cities as post-industrial places has been about shoring up the authority of local authorities. These reinventions invariably rewrite identity in less progressive ways, silencing the presence and history of working-class, ethnic minorities, as well as other aspects of the local landscape considered less saleable (Chapters 4 and 6). Dispossessing minority groups of their land, such as indigenous people in settler countries, is also a political strategy, fundamentally and permanently disempowering these groups. We discussed in Chapter 4 how shopping malls are constructed in a manner to control behaviour. The myths of elsewhereness within these landscapes are intended to will a suspension from reality and lull shoppers into an expanded expenditure of time and money. The Taliban regime in Afghanistan legislated and violently policed the corporeality of women. Vigorously scrutinising these landscapes was a means for the patriarchal control of women (Chapter 7). All societies have standards against which bodies are deemed to be deviant or normal. In western societies women who are pregnant find that their behaviour is policed by their peers and in some case by courts (Chapter 7). Clearly, the cultural landscape is used by powerful groups to shore up and, if possible, naturalise uneven power relations.

But power is multiplicitious and less powerful groups and individuals are just as likely to utilise the landscape for political ends. Constructing the cultural landscape in certain ways allows minority groups to make claims to space. Murals and graffiti can have this aim and effect (Chapters 3 and 5). The excluded themselves can claim residual landscapes, which they can construct as liberated zones, islands in a sea of racism, homophobia, sexism and such like. Landscapes are a resource – they are a means of cultural reproduction and a base for political action. These roles were apparent in many case studies, including the so-called gay precincts of Castro (San Francisco) and Oxford Street (Sydney) (Chapter 5). Reclaim the Night marches are attempts to claim non-residual public spaces. The participants in these marches take their protest to the naturalised sexism of everyday landscapes (Chapter 5). Local communities have been able to argue successfully that McDonald's Family Restaurants are out of place in their landscapes (Chapter 3). Some landscapes of resistance are more ephemeral in nature. Carnivals and mardi gras are examples of temporary landscapes of transgression and rebellion to norms. The events reviewed in Chapter 5 temporarily queered city streets, asserted multiculturalism, projected non-domestic femininities and challenged the authority of police. Bodies can also be landscapes of resistance. Feminists have used dress to challenge patriarchal notions of femininity and youth have used scarring, tattoos and clothing to inscribe their rejection of the authority of elders and elite (Chapter 7). The so-called 'lipstick lesbian' both confronts patriarchal assumptions about femininity and rebels against orthodoxies of dress that had developed within lesbian politics (Chapter 7). Clearly, there are landscapes of both oppression and resistance.

8.3 The contribution of *Landscapes* _____

A wealth of case study material, much of it developed by cultural geographers, has been presented in this book. This material supports the many points made about cultural landscapes and the roles they have. What this evidence demonstrates in totality is the societal importance of cultural landscapes. Cultural landscapes, and their representations, are clearly worthy of sustained intellectual analysis. Our fervent hope is that we have also provided readers with the basic analytical tools for analysing cultural landscapes.

References

Abbey, R. and Crawford, J. (1987) 'Crocodile Dundee or Davy Crockett? What *Crocodile Dundee* doesn't say about Australia', *Meanjin*, 46(2), 145–52.

Abel, A. (2000) 'The art of vandalism', *Saturday Night*, 115(2), 52–9.

Adams, P. (1986) 'Sorry Hoges, but this time you've blown it', *The Weekend Australian*, 26 April, M2.

Adler, S. and Brenner, J. (1992) 'Gender and space: lesbians and gay men in the city', *International Journal of Urban and Regional Research*, 16, 24–34.

Adorno, T. (1992) *Quasi Una Fantasia*, translated by R. Livingstone, London: Verso.

Agnew, J.A. (1987) *Place and Politics: The Geographical Mediation of State and Society*, London: Allen & Unwin.

Agnew, J.A. and Corbridge, S. (1995) *Mastering Space: Hegemony, Territory and International Political Economy*, London: Routledge.

Agnew, J.A., Mercer, J. and Sopher, D.E. (1984) 'Introduction', in *The City in Cultural Context*, Agnew, J.A., Mercer, J. and Sopher, D.E. (eds), Boston: Allen & Unwin, 1–30.

Ahmad, N. (1947) *Muslim Contribution to Geography*, Lahore: Muhammed Ashraf.

Aitken, S.C. and Zonn, L.E. (1993) 'Weir(d) sex: representation of gender-environment relations in Peter Weir's *Picnic at Hanging Rock* and *Gallipoli*', *Environment and Planning D: Society and Space*, 11(2), 191–212.

Akama, J.S. (1996) 'Western environmental values and nature-based tourism in Kenya', *Tourism Management*, 17(8), 567–74.

Almgren, H. (1994) 'Community with/out pro-pink-uity', in *The Margins of the City: Gay Men's Urban Lives*, Whittle, S. (ed.), Aldershot: Arena, 45–63.

Anderson, B. (1972) 'The idea of power in Javanese culture', in *Culture and Politics in Indonesia*, Holt, C. (ed.), Ithaca: Cornell University Press.

Anderson, K.J. (1987) 'The idea of Chinatown: the power of place and institutional practice in the making of a racial category', *Annals of the Association of American Geographers*, 77(4), 580–98.

Anderson, K.J. (1988) 'Cultural hegemony and the race-definition process in Chinatown, Vancouver: 1880–1980', *Environment and Planning D: Society and Space*, 6(2), 127–49.

Anderson, K.J. (1993) 'Otherness, culture and capital: "Chinatown's" transformation under Australian multiculturalism', in *Multiculturalism, Difference and Postmodernism*, Clark, G.L., Forbes, D. and Francis, R. (eds), Melbourne: Longman Cheshire, 68–89.

Appadurai, A. (1990) 'Disjuncture and difference in the global cultural economy', in *Global Culture, Nationalism, Globalization and Modernity*, Featherstone, M. (ed.), London: Sage, 295–310.

Astarotte, A. (2001) 'The geography of hell and purgatory: some brief notes', *IAG Newsletter*, 44 (June), 22–6.

Attwood, A. and Helley, K. (1988) 'Taking Mick out of an Aussie myth', *Time* (Australia), 30 May, 72.

Australian Film Commission (1998) *Getting the Picture: Essential Data on Australia Film, Television, Video and New Media*, Sydney: Australian Film Commission.

Azaryahu, M. (1992) 'The purge of Bismarck and Saladin: the renaming of streets in East Berlin and Haifa: a comparative study in culture-planning', *Poetics Today*, 13, 351–67.

Azaryahu, M. (1996) 'The power of commemorative street names', *Environment and Planning D: Society and Space*, 14(3), 311–30.

Badcock, B. (1996) '"Looking-glass" views of the city', *Progress in Human Geography*, 20(1), 91–9.

Bakhtin, M. (1968) *Rabelais and his World*, Cambridge, MA: MIT Press.

Barthes, R. (1972) *Mythologies*, selected and translated from the French by A. Lavers, London: Jonathan Cape.

Bell, D. (1991) 'Insignificant others: lesbian and gay geographies', *Area*, 23(4), 323–9.

Bell, D. and Valentine, G. (1997) *Consuming Geographies: We are Where We Eat*, London: Routledge.

Bell, D., Binnie, J., Crean, J. and Valentine, G. (1994) 'All hyped up and no place to go', *Gender, Place and Culture*, 1(1), 31–47.

Benedict, B. (1983) 'Sport and cultural reification: from ritual to mass consumption', *Theory, Culture and Society*, 1, 93–107.

Bennett, S. (1997) *The Social Construction of a Bounded Safe Space for Sexual Dissidents and its Impact on Anti-gay and Lesbian Violence*, BSc Honours Thesis, School of Geography, Sydney: University of New South Wales.

Benton, L.M. (1995) 'Will the real/reel Los Angeles please stand up?', *Urban Geography*, 16(2), 144–64.

Bergman, E.F. (1995) *Human Geography: Cultures, Connections, and Landscapes*, Englewood Cliffs: Prentice-Hall.

Bevan, C. (1991) 'Brass band contests: art or sport?', in *Bands: The Brass Band Movement in the 19th and 20th Centuries*, Herbert, T. (ed.), Milton Keynes: Open University Press.

Blagg, T. (1986) 'Roman religious sites in the British landscape', *Landscape History*, 8, 15–26.

Blake, P. (1972) 'Walt Disney world', *Architectural Forum*, 136, 24–41.

Blakely, E.J. and Snyder, M.G. (1998) 'Forting up: gated communities in the United States', *Journal of Architectural and Planning Research*, 15(1), 61–72.

Boles, D. (1989) 'Reordering the suburbs', *Progressive Architecture*, 70, 78–91.

Bonnett, A. (1996) 'Constructions of "race", place and discipline: geographies of "racial" identity and racism', *Ethnic and Racial Studies*, 19(4), 864–83.

Bowen, T. (1999) 'Graffiti art: a contemporary study of Toronto artists', *Studies in Art Education*, 41(1), 22–30.

Bowlby, S., Lewis, J., McDowell, L. and Foord, J. (1989) 'The geography of gender', in *New Models in Geography: The Political-Economy Perspective*, Peet, R. and Thrift, N. (eds), London: Unwin Hyman, 157–75.

Bradshaw, M. and Williams, S. (1999) 'Scales, lines and minor geographies: whither King Island?', *Australian Geographical Studies*, 37(3), 248–67.

Brumberg, J. (1988) *Fasting Girls: The Emergence of Anorexia Nervosa as a Modern Disease*, Cambridge: Harvard University Press.

Buckley, C.B. (1984) *An Anecdotal History of Old Times in Singapore*, Singapore: Oxford University Press.

Burgess, J. (1985) 'News from nowhere: the press, the riots and the myth of the inner city', in *Geography, the Media and Popular Culture*, Burgess, J. and Gold, J.R. (eds), Kent: Croom helm, 192–228.

Burgess, J. (1990) 'The production and consumption of environmental meanings in the mass media: a research agenda for the 1990s', *Transactions of the Institute of British Geographers*, 15(2), 139–61.

Burgess J. and Gold, J.R. (eds) (1985) *Geography, the Media and Popular Culture*, Kent: Croom Helm.

Butler, R. and Bowlby, S. (1997) 'Bodies and spaces: an exploration of disabled people's experiences of public space', *Environment and Planning D: Society and Space*, 15(4), 411–33.

Bynum, C. (1987) *Holy Feast and Holy Fast: The Religious Significance of Food to Medieval Women*, Berkeley: University of California Press.

Carney, G. (1990) 'Geography of music: inventory and prospect', *Journal of Cultural Geography*, 10(2), 35–48.

Carstensen, L. (1995) 'The burger kingdom: growth and diffusion of McDonald's restaurants in the United States, 1955–1978', in *Fast Food, Stock Cars, and Rock 'n' Roll: Place and Space in American Popular Culture*, Carney, G.O. (ed.), Lanham: Rowman & Littlefield, 119–30.

Carter, D. (1996) 'Crocs in frocks: landscape and nation in the 1990s', *Journal of Australian Studies*, 49, 89–96.

Castells, M. (1977) *The Urban Question*, London: Edward Arnold.

Castells, M. (1983) *The City and the Grassroots: A Cross-Cultural Theory of Urban Social Movements*, London: Edward Arnold.

Celik, Z. (1992) *Displaying the Orient: Architecture of Islam at 19th Century World's Fairs*, Berkeley: University of California Press.

Chalmers, M. (1988) 'Heroin, the needle and the politics of the body', in *Zoot Suits and Second-hand Dresses: An Anthology of Fashion and Music*, McRobbie, A. (ed.), Boston: Unwin Hyman, 150–55.

Chouinard, V. and Grant, A. (1995) 'On being not even anywhere near "The Project": ways of putting ourselves in the picture', *Antipode*, 27(2), 137–66.

Chua, B.H. (1991) 'Modernism and the vernacular: transformation of public spaces and social life in Singapore', *Journal of Architectural and Planning Research*, 8, 203–21.

Clifford, J. (1986) 'Partial truths', in *Writing Culture: The Poetics and Politics of Ethnography*, Clifford, J. and Marcus, G.E. (eds), Berkeley: University of California Press, 1–26.

Cohen, A.P. (1982) 'Belonging: the experience of culture', in *Belonging: Identity and Social Organization in British Rural Cultures*, Cohen, A.P. (ed.), Manchester: Manchester University Press, 1–17.

Cohen, S. (1973) *Folk Devils and Moral Panics*, St Albans: Granada Publishing.

Cohen, S. (1991) 'Popular music and urban regeneration: the music industries of Merseyside', *Cultural Studies*, 5, 332–46.

Collmann, J. (1988) *Fringe-dwellers and Welfare: The Aboriginal Response to Bureaucracy*, St Lucia: University of Queensland Press.

Colombijn, F. (1998) 'A sheep in wolf's clothing', *International Journal of Urban and Regional Research*, 22(4), 565–81.

Connell, R.W. (1995) *Masculinities*, Berkeley: University of California Press. *Conservation within the Central Area with the Plan for Chinatown* (1985) Singapore: Urban Redevelopment Authority, 15.

Cosgrove, D. (1982) 'The myth and the stones of Venice: an historical geography of a symbolic landscape', *Journal of Historical Geography*, 8(2), 145–69.

Cosgrove, D. (1989) 'A terrain of metaphor: cultural geography 1988–89', *Progress in Human Geography*, 13(4), 566–75.

Cosgrove, D. and Jackson, P. (1987) 'New directions in cultural geography', *Area*, 19, 95–101.

Costello, L.N. and Dunn, K.M. (1994) 'Resident action groups in Sydney: people power or rat-bags?', *Australian Geographer*, 25(1), 61–76.

Cowlishaw, G. (1988) 'The materials for identity construction', in *Past and Present: The Construction of Aboriginality*, Beckett, J. (ed.), Canberra: Aboriginal Studies Press, 87–108.

Cresswell, T. (1992) 'The crucial "where" of graffiti: a geographical analysis of re-actions to graffiti in New York', *Environment and Planning D: Society and Space*, 10(3), 329–44.

Cresswell, T. (1994) 'Putting women in their place: the Carnival at Greenham Common', *Antipode*, 26(1), 35–58.

Crofts, S. (1989) 'Re-imaging Australia: Crocodile Dundee overseas', *Continuum*, 2(2), 129–42.

Crofts, S. (1992) 'Cross-cultural reception studies: culturally variant readings of *Crocodile Dundee*', *Continuum*, 6(1), 213–27.

Cunneen, C. and Lynch, R. (1988) 'The social meanings of conflict in riots at the Australian Grand Prix motorcycle races', *Leisure Studies*, 7(1), 1–19.

Cunneen, C., Findlay, M., Lynch, R. and Tupper, V. (1989) *Dynamics of Collective Conflict: Riots at the Bathurst Bike Races*, Sydney: Law Book.

Cuthbert, A.R. (1995) 'The right to the city: surveillance, private interest and the public domain in Hong Kong', *Cities*, 12(5), 293–310.

Cuthbert, A.R. and McKinnell, K.G. (1997) 'Ambiguous space, ambiguous rights – corporate power and social control in Hong Kong', *Cities*, 14(5), 295–311.

Cybriwsky, R., Ley, D. and Western, J. (1986) 'The political and social construction of revitalized neighborhoods: Society Hill, Philadelphia, and False Creek, Vancouver', in *Gentrification of the City*, Smith, N. and Williams, P. (eds), London: Unwin Hyman, 92–120.

Dahlberg, A. (1991) 'The body as a principle of holism: three pilgrimages to Lourdes', in *Contesting the Sacred: The Anthropology of Christian Pilgrimage*, Eade, J. and Sallnow, M. (eds), London: Routledge, 30–50.

Daniels, S. (1993) *Fields of Vision: Landscape Imagery and National Identity in England and the United States*, Cambridge: Polity Press.

Davidson, J. (1987) 'Locating *Crocodile Dundee*', *Meanjin*, 46(1), 122–8.

Davis, L.J. (1995) *Enforcing Normalcy: Disability, Deafness, and the Body*, London: Verso.

Davis, M. (1990) *City of Quartz: Excavating the Future in Los Angeles*, London: Verso.

Davis, S.G. (1996) 'The theme park: global industry and cultural form', *Media, Culture and Society*, 18, 399–422.

Davis, T. (1995) 'The diversity of queer politics and the redefinition of sexual identity and community in urban spaces', in *Mapping Desire: Geographies of Sexualities*, Bell, D. and Valentine, G. (eds), London: Routledge, 284–303.

Day, G. (1990) 'Pose for thought: body building and other matters', in *Readings in Popular Culture*, Day, G. (ed.), Basingstoke: Macmillan, 48–58.

Day, P.D. (1986) 'Canberra: best planned or most planned?', *Urban Policy and Research*, 4(4), 14–21.

de Blij, H.J. (1977) *Human Geography: Culture, Society, and Space*, New York: John Wiley.

de Certeau, M. (1984) *The Practice of Everyday Life*, translated by S. Rendall, Berkeley: University of California Press.

Dear, M.J. and Wolch, J.R. (1987) *Landscapes of Despair: From Deinstitutionalization to Homelessness*, Oxford: Polity Press.

Dear, M.J., Gaber, S., Takahashi, L. and Wilton, R. (1997) 'Seeing people differently: the sociospatial construction of disability', *Environment and Planning D: Society and Space*, 14(4), 455–77.

Deffontaines, P. (1953) 'The religious factor in human geography', *Diogenes*, 2, 24–37.

Diamond, J. (1997) *Guns, Germs and Steel: A Short History of Everybody for the Last 13,000 Years*, London: Vintage Books.

Dickens, C. (1949) *The Adventures of Oliver Twist*, London: Oxford University Press.

Dickenson, J.P., Clarke, C.G., Gould, W.T.S., Prothero, R.M., Siddle, D.J., Smith, C.T., Thomas-Hope, E.M. and Hodgkiss, A.G. (1983) *A Geography of the Third World*, London: Routledge.

Dickinson, R.E. (1939) 'Landscape and society', *Scottish Geographical Magazine*, 55, 1–15.

Domosh, M. (1989) 'New York's first skyscrapers: conflict in design of the American commercial landscape', *Landscape*, 30(2), 34–8.

Doogue, E. (1987) 'Dundee's just a New Right bully, says Sister Veronica', *The Age*, 31 January, 3.

Dorn, M. and Laws, G. (1994) 'Social theory, body politics and medical geography', *The Professional Geographer*, 46(1), 106–10.

Driver, F. (1989) 'The historical geography of the workhouse system in England and Wales, 1834–1883', *Journal of Historical Geography*, 15, 269–86.

Duffy, M. (1994) 'Suburbia's new cathedrals', *Independent Monthly* (March), 28–33.

Duggan, L. (1994) 'Queering the state', *Social Text*, 39 (Summer), 1–24.

Duncan, J.S. (1980) 'The superorganic in American cultural geography', *Annals of the Association of American Geographers*, 70(2), 181–98.

Duncan, J.S. (1990) *The City as Text: The Politics of Landscape Interpretation in the Kandyan Kingdom*, Cambridge: Cambridge University Press.

Duncan, J.S. (1994) 'The politics of landscape and nature, 1992–3', *Progress in Human Geography*, 18(3), 361–70.

Duncan, J.S. and Duncan, N. (1988) '(Re)reading the landscape', *Environment and Planning D: Society and Space*, 6(2), 117–26.

Duncan, J.S. and Ley, D. (eds) (1993) *Place/Culture/Representation*, London: Routledge.

Dunn, K.M. (1993) 'The Vietnamese concentration in Cabramatta: site of avoidance and deprivation, or island of adjustment and participation?', *Australian Geographical Studies*, 31(2), 228–45.

Dunn, K.M. (1997) 'Cultural geography and cultural policy', *Australian Geographical Studies*, 35(1), 1–11.

Dunn, K.M. (2001) 'Representations of Islam in the politics of mosque development in Sydney', *Tijdschrift voor Economische en Sociale Geografie*, 92(3), 291–308.

Dunn, K.M. and Mahtani, M. (2001) 'Media representations of ethnic minorities', *Progress in Planning*, 53(3), 163–72.

Dunn, K.M. and McGuirk, P.M. (1999) 'Hallmark events', in *Staging the Olympics – the Event and its Impact*, Cashman, R. and Hughes, A. (eds), Sydney: UNSW Press, 19–32.

Dunn, K.M. and Roberts, S. (forthcoming) 'The social construction of an Indo-Chinese-Australian neighbourhood in Sydney: the case of Cabramatta', in *From Urban Enclave to Ethnic Suburb: New Asian Communities in Pacific Rim Countries*, Li, W. (ed.), Baltimore: John Hopkins University Press.

Dunn, K.M. and Winchester, H.P.M. (1999) 'Invention of gender and place in film: tales of urban reality', in *Cultural Geographies*, Anderson, K. and Gale, F. (eds), 2nd edn, Melbourne: Addison Wesley Longman, 173–95.

Dunn, K.M., McGuirk, P.M. and Winchester, H.P.M. (1995) 'Place making: the social construction of Newcastle', *Australian Geographical Studies*, 33(2), 149–66.

During, S. (1993) 'Introduction', in *The Cultural Studies Reader*, During, S. (ed.), London: Routledge, 1–25.

Durkheim, E. (1964) *The Division of Labour in Society*, New York: Free Press.

Dwyer, C. (1999) 'Contradictions of community: questions of identity for young British Muslim women', *Environment and Planning A*, 31(1), 53–68.

El Erian, M.A. (1990) *Jamjoom: A Profile of Islam, Past, Present, Future*, Melbourne: Islamic Publications.

Enjeu, C. and Save, J. (1974) 'The city: off limits to women', *Liberation*, 18, 9–13.

Featherstone, M. (1993) 'Global and local cultures', in *Mapping the Futures: Local Cultures, Global Change*, Bird, J., Curtis, B., Putnam, T., Robertston, G. and Tickner, L. (eds), London: Routledge, 169–87.

Fellmann, J., Getis, A. and Getis, J. (1995) *Human Geography: Landscapes of Human Activity*, Dubuque: WMC Brown.

Ferrell, J. (1995) 'Urban graffiti: crime, control, and resistance', *Youth and Society*, 27(1), 73–92.

Fickeler, P. (1962) 'Fundamental questions in the geography of religions', in *Readings in Cultural Geography*, Wagner, P.L. and Mikesell, M.W. (eds), Chicago: University of Chicago Press, 94–117.

Finnegan, R. (1989) *The Hidden Musicians: Music-making in an English Town*, Cambridge: Cambridge University Press.

Fishwick, M. (1995) 'Ray and Ronald girdle the globe', *Journal of American Popular Culture*, 18(1), 13–29.

Fiske, J. (1989) *Reading the Popular*, Boston: Unwin Hyman, 43–76.

Flannery, T. (1994) *The Future Eaters: An Ecological History of the Australasian Lands and People*, Sydney: Reed Books.

Flood, J. (1995) *Archaeology of the Dreamtime, the Story of Prehistoric Australia and its People*, Sydney: Angus & Robertson.

Folch-Serra, M. (1990) 'Place, voice, space: Mikhail Bakhtin's dialogical-landscape', *Environment and Planning D: Society and Space*, 8(3), 255–74.

Foucault, M. (1979) *Discipline and Punish: The Birth of the Prison*, translated by A. Sheridan, London: Penguin.

Foucault, M. (1980a) 'The eye of power', in *Michel Foucault: Power/Knowledge, Selected Interviews and Other Writings, 1972–1977*, Gordon, C. (ed.), New York: Harvester Press, 146–65.

Foucault, M. (1980b) 'Power and strategies', in *Michel Foucault: Power/Knowledge, Selected Interviews and Other Writings, 1972–1977*, Gordon, C. (ed.), New York: Harvester Press, 134–45.

Foucault, M. (1980c) 'Two lectures', in *Michel Foucault: Power/Knowledge, Selected Interviews and Other Writings, 1972–1977*, Gordon, C. (ed.), New York: Harvester Press, 78–108.

Foucault, M. (1984) 'Truth and power', in *The Foucault Reader*, Rabinow, P. (ed.), Harmondsworth: Penguin, 51–75.

Frith, S. (ed.) (1989) *World Music, Politics and Social Change*, Manchester: Manchester University Press and International Association for the Study of Popular Music.

Frow, J. and Morris, M. (1993) 'Introduction', in *Australian Cultural Studies: A Reader*, Frow, J. and Morris, M. (eds), St Leonards: Allen & Unwin, vii–xxxii.

Furth, C. (1987) 'Concepts of pregnancy, childbirth, and infancy in Ch'ing Dynasty China', *Journal of Asian Studies*, 46(1), 7–35.

Gale, F. (1972) *Urban Aborigines*, Canberra: Australian National University Press.

Gatens, M. (1991) 'A critique of the sex/gender distinction', in *A Reader in Feminist Knowledge*, Gunew, S. (ed.), London: Routledge, 139–57.

Gay, J.D. (1971) *The Geography of Religion in England*, London: Duckworth.

Gimlin, D. (1996) 'Pamela's place: power and negotiation in the hair salon', *Gender and Society*, 10(5), 505–26.

Gleeson, B. (1995) 'Disability: a state of mind?', *Australian Journal of Social Issues*, 30(1), 10–23.

Gleeson, B. (1999) 'Recovering a "subjugated history": disability and the institution in the industrial city', *Australian Geographical Studies*, 37(2), 114–29.

Goetz, Edward, G. (1992) 'Land use and homeless policy in Los Angeles', *International Journal of Urban and Regional Research*, 16(4), 540–54.

Gold, J.R. (1985) 'From *Metropolis* to *The City*: film visions of the future city, 1919–39', in *Geography, the Media and Popular Culture*, Burgess, J. and Gold, J.R. (eds), Kent: Croom Helm, 123–43.

Goldberg, J. (2000) 'The spectre of McDonald's: an object of bottomless hatred', *National Review*, 5 June, 29–32.

Golledge, R. (1997) 'On reassembling one's life: overcoming disability in the academic environment', *Environment and Planning D: Society and Space*, 15(4), 391–409.

Goodall, H., Jakubowicz, A., Martin, J., Mitchell, T., Randall, L. and Seneviratne, K. (1994) *Racism, Ethnicity and the Media*, St Leonards: Allen & Unwin.

Gordon, C. (1980) 'Afterword', in *Michel Foucault: Power/Knowledge, Selected Interviews and Other Writings, 1972–1977*, Gordon, C. (ed.), New York: Harvester Press, 229–59.

Goss, J. (1988) 'The built environment and social theory: towards an architectural geography', *The Professional Geographer*, 40(4), 392–403.

Goss, J. (1993) 'The magic of the mall: an analysis of form, function, and meaning in the contemporary retail built environment', *Annals of the Association of American Geographers*, 83(1), 18–47.

Goss, J. (1999) 'Once-upon-a-time in the commodity world: an unofficial guide to the mall of America', *Annals of the Association of American Geographers*, 89(1), 45–75.

Graham, B.J. (1994) 'No place of the mind: contested Protestant representations of Ulster', *Ecumene*, 1(3), 257–81.

Gramsci, A. (1973) *Letters from Prison*, New York: Harper & Row.

Greenberg, M., Popper, F., West, B. and Schneider, D. (1992) 'TOADs go to New

Jersey: implications for land use and public health in medium-sized and large US cities', *Urban Studies*, 29(1), 117–25.

Greenhalgh, P. (1988) *Ephemeral Vistas: The Expositions Universelles, Great Exhibitions and World's Fairs, 1851–1939*, New York: Manchester University Press.

Gregson, N. (1993) 'The 1988 initiative: delimiting or deconstructing social geography?', *Progress in Human Geography*, 17(4), 525–30.

Gross, D.D., Walkosz, B. and Gross, T.D. (1997) 'Language boundaries and discourse stability', *Et cetera*, 54, 275–85.

Grosz, E. (1987) 'Notes towards a corporeal feminism', *Australian Feminist Studies*, 5 (Summer), 1–16.

Hall, G.B. and Joseph, A.E. (1988) 'Group home location and host neighborhood attributes: an ecological analysis', *The Professional Geographer*, 40(3), 297–306.

Hall, S. (1981) 'Notes on deconstructing the "popular"', in *People's History and Socialist Theory*, Samuel, R. (ed.), London: Routledge & Kegan Paul, 227–40.

Hall, S. (1990) 'The emergence of cultural studies and the crisis of humanities', *October*, 53, 11–23.

Hall, S. and Jefferson, T. (eds) (1976) *Resistance Through Rituals: Youth Subcultures in Post-war Britain*, London: Hutchinson.

Hall, S., Critcher, C. and Jefferson, T. (1978) *Policing the Crisis: Mugging, the State and Law and Order*, London: Macmillan.

Hamilton, A. (1990) 'Fear and desire: Aborigines, Asians and the national imaginary', *Australian Cultural History*, 9, 14–35.

Haraway, D. (1990) *Simians, Cyborgs and Women: The Reinvention of Nature*, London: Free Association Books.

Hartig, K.V. and Dunn, K.M. (1998) 'Roadside memorials: interpreting new deathscapes in Newcastle, New South Wales', *Australian Geographical Studies*, 36(1), 5–20.

Hartshorne, R. (1939) 'The nature of geography: a critical survey of current thought in the light of the past', *Annals of the Association of American Geographers*, 29(3–4).

Harvey, D. (1979) 'Monument and myth', *Annals of the Association of American Geographers*, 69(3), 362–81.

Harvey, D. (1992) 'Social justice, postmodernism and the city', *International Journal of Urban and Regional Research*, 16(4), 588–601.

Harvey, D. (1996) *Justice, Nature and the Geography of Difference*, Cambridge, Massachusetts: Blackwell.

Hastings, A. (1999) 'Discourse and urban change: introduction to the special issue', *Urban Studies*, 36(1), 7–12.

Hatch, M. (1989) 'Popular music in Indonesia', in *World Music, Politics and Social Change*, Frith, S. (ed.), Manchester: Manchester University Press, 47–67.

Henzel, C. (1989) '*Cruces* in the roadside landscape of northeastern Mexico', *Journal of Cultural Geography*, 11, 93–106.

Heppel, M. (ed.) (1979) *A Black Reality: Aboriginal Camps and Housing in Remote Australia*, Canberra: Australian Institute of Aboriginal Studies.

Herbert, T. (1992) 'Victorian brass bands: establishment of a working class musical tradition', *Historic Brass Society Journal*, 4, 1–11.

Herbertson, A.J. (1905) 'The major natural regions: an essay in systematic geography', *Geography Journal*, 25, 300–12.

Higgins, W. (1982) 'To him that hath . . . : the welfare state', in *Australian Welfare History: Critical Essays*, Kennedy, R. (ed.), Melbourne: Macmillan, 199–224.

Hillier, J. and McManus, P. (1994) 'Pull up the drawbridge: fortress mentality in the suburbs', in *Metropolis Now: Planning and the Urban in Contemporary Australia*, Gibson, K. and Watson, S. (eds), Sydney: Pluto Press, 91–107.

Hindle, P. (1994) 'Gay communities and gay space in the city', in *The Margins of the City: Gay Men's Urban Lives*, Whittle, S. (ed.), Aldershot: Arena, 7–25.

Hodge, S. (1995) '"No fags out there": gay men, identity and suburbia', *Journal of Interdisciplinary Gender Studies*, 1(1), 41–8.

Hoggart, R. (1957) *The Uses of Literacy*, Harmondsworth: Penguin.

Homes for the People: A Review of Public Housing by the Singapore Housing Development Board (no date) Singapore: The Straits Times Press.

Hopkins, J.S.P. (1990) 'West Edmonton Mall: landscape of myths and elsewhereness', *The Canadian Geographer*, 34(1), 2–17.

Housing and Development Board Annual Report (1961), Singapore.

Housing and Development Board Annual Report (1969), Singapore.

Hsu, S.-Y. (1969) 'The cultural ecology of the Locust Cult in traditional China', *Annals of the Association of American Geographers*, 59, 730–52.

Hubbard, P. (1998) 'Community action and the displacement of street prostitution: evidence from British cities', *Geoforum*, 29(3), 269–86.

Human Genome Diversity Committee (1993) *Human Genome Diversity Project Outline*, Stanford: Stanford University Press.

Hunt, J.D. (2000) *Greater Perfections: The Practice of Garden Theory*, Philadelphia: University of Pennsylvania Press.

Huxley, M. and Kerkin, K. (1988) 'What price the Bicentennial? A political economy of Darling Harbour', *Transition: Discourse on Architecture*, 26, 57–64.

Ibn Khordathanak (846) *Book of Routes and Kingdoms*.

Isaac, E. (1959–60) 'Religion, landscape and space', *Landscape*, 9(2), 14–18.

Jackson, P. (1988) 'Street life: the politics of Carnival', *Environment and Planning D: Society and Space*, 6, 213–27.

Jackson, P. (1989) *Maps of Meaning: An Introduction to Cultural Geography*, London: Unwin Hyman.

Jackson, P. (1991) 'The cultural politics of masculinity: towards a social geography', *Transactions of the Institute of British Geographers*, 16(2), 199–213.

Jackson, P. and Penrose, J. (eds) (1993) *Constructions of Race, Place and Nation*, London: University College London Press.

Jacobs, K. (1999) 'Key themes and future prospects: conclusion to the special issue', *Urban Studies*, 36(1), 203–13.

Jameson, F. (1984) 'Foreword', in *The Postmodern Condition: A Report on Knowledge*, Lyotard, J., Minneapolis: University of Minnesota Press, vii–xxi.

Jameson, F. (1988) 'Postmodernism and consumer society', in *Postmodernism and its Discontents: Theories, Practices*, Kaplan, E.A. (ed.), London: Verso, 13–29.

Jarman, N. (1998) 'Painting landscapes: the place of murals in the symbolic construction of urban space', in *Symbols in Northern Ireland*, Buckley, A. (ed.), Belfast: Institute of Irish Studies, Queen's University of Belfast.

Jeans, D.N. (1988) 'The First World War memorials in New South Wales: centres of meaning in the landscape', *Australian Geographer*, 19(2), 259–67.

Jones, M.L. (ed.) (1993) *An Australian Pilgrimage*, Melbourne: Victoria Press.

Jones, R. (1969) 'Fire-stick farming', *Australian Natural History*, 9, 224–8.

Jones, R. (1997) 'Sacred sites or profane buildings? Reflections on the Old Swan Brewery conflict in Perth, Western Australia', in *Contested Urban Heritage: Voices from the Periphery*, Shaw, B. and Jones, R. (eds), Aldershot: Ashgate, 132–55.

Jordan, T.G. (1973) *The European Culture Area*, New York: Harper & Row.

Kan, K.-H. (2001) 'Adolescents and graffiti', *Art Education*, 54(1), 18–23.

Kennedy, C. and Luckinbeal, C. (1997) 'Towards a holistic approach to geographic research on film', *Progress in Human Geography*, 21(1), 33–50.

King, A.D. (1976) *Colonial Urban Development: Culture, Social Power, and Environment*, London: Routledge & Kegan Paul.

Kirby, A. (1995) 'Straight talk on the pomohomo question', *Gender, Place and Culture*, 2(1), 89–95.

Kirby, S. and Hay, I. (1997) '(Hetero)sexing space: gay men and "straight" space in Adelaide, South Australia', *The Professional Geographer*, 49(3), 296–310.

Klingman, A., Shalev, R. and Pearlman, A. (2000) 'Graffiti: a creative means of youth coping with collective trauma', *The Arts in Psychotherapy*, 27(5), 299–307.

Kobayashi, A. and Peake, L. (1994) 'Unnatural discourse: "race" and gender in geography', *Gender, Place and Culture*, 1, 225–43.

Kong, L. (1990) 'Geography and religion: trends and prospects', *Progress in Human Geography*, 14(3), 355–71.

Kong, L. (1993) 'Ideological hegemony and the political symbolism of religious buildings in Singapore', *Environment and Planning D: Society and Space*, 11(1), 23–45.

Kong, L. (1995) 'Music and cultural politics: ideology and resistance in Singapore', *Transactions, Institute of British Geographers*, 20(4), 447–59.

Kong, L. (1996a) 'Popular music in Singapore: local cultures, global resources and regional identities', *Environment and Planning D: Society and Space*, 14(3), 273–92.

Kong, L. (1996b) 'Making "music at the margins"? A social and cultural analysis of xinyao in Singapore', *Asian Studies Review*, 19, 99–124.

Kong, L. (1997) 'A "new" cultural geography? Debates about invention and reinvention', *Scottish Geographical Magazine*, 113(3), 177–85.

Kong, L. (1999) 'Cemeteries and columbaria, memorials and mausoleums: narrative and interpretation in the study of deathscapes in geography', *Australian Geographical Studies*, 37(1), 1–10.

Kong, L. (2001) 'Mapping "new" geographies of religion: politics and poetics of modernity', *Progress in Human Geography*, 25(2), 211–33.

Kong, L. and Yeoh, B.S.A. (1996) 'Social constructions of nature in urban Singapore', *Southeast Asian Studies*, 34, 402–23.

Kong, L. and Yeoh, B.S.A. (2002) *Constructions of 'Nation': The Politics of Landscapes in Singapore*, Syracuse: Syracuse University Press.

Kristeva, J. (1980) 'Motherhood according to Giovanni Bellini', in *Desire in Language: A Semiotic Approach to Literature and Art*, Roudiez, L. (ed.), translated by A. Jardine, T. Gora and L. Roudiez, New York: Columbia University Press, 237–70.

Kristeva, J. (1982) *Powers of Horror: An Essay on Abjection*, New York: Columbia University Press.

Lachmann, R. (1988) 'Graffiti as career ideology', *American Journal of Sociology*, 94(2), 229–50.

Lash, S. (1990) *Sociology of Postmodernism*, London: Routledge.

Law, L. (2000) *Sex Work in Southeast Asia: The Place of Desire in a Time of AIDS*, London: Routledge.

Law, L. (2001) 'Home cooking: Filipino women and geographies of the senses in Hong Kong', *Ecumene*, 8(3), 264–83.

Lawrence, D. (1982) 'Parades, politics and competing urban images: Doo Dah and Roses', *Urban Anthropology*, 4(2), 155–76.

Laws, G. (1994) 'Oppression, knowledge and the built environment', *Political Geography*, 13(1), 7–32.

Lederer, L. (ed.) (1980) *Take Back the Night: Women on Pornography*, New York: William Morrow.

Lefebvre, H. (1991) *The Production of Space*, translated by D. Nicholson-Smith, Oxford: Blackwell.

Lewis, C. and Pile, S. (1996) 'Woman, body, space: Rio Carnival and the politics of performance', *Gender, Place and Culture*, 3(1), 23–42.

Lewis, P.F. (1979) 'Axioms for reading the landscape', in *The Interpretation of Ordinary Landscapes: Geographical Essays*, Meinig, D.W. (ed.), New York: Oxford University Press, 11–32.

Ley, D. (1981) 'Inner city revitalization in Canada: a Vancouver case study', *The Canadian Geographer*, 25, 124–48.

Ley, D. and Cybriwsky, R. (1974) 'Urban graffiti as territorial markers', *Annals of the Association of American Geographers*, 64(4), 491–505.

Ley, D. and Olds, K. (1988) 'Landscape as spectacle: world's fairs and the culture of heroic consumption', *Environment and Planning D: Society and Space*, 6(2), 191–212.

Ley, D. and Olds, K. (1999) 'World's fairs and the culture of consumption in the contemporary city', in *Cultural Geographies*, Anderson, K. and Gale, F. (eds), 2nd edn, Melbourne: Longman, 221–40.

Longhurst, R. (1994) 'The geography closest in – the body . . . the politics of pregnability', *Australian Geographical Studies*, 32(2), 214–23.

Loyd, B. (1975) 'Women's place, man's place', *Landscape*, 20, 10–13.

Loyd, B. and Rowntree, L. (1978) 'Radical feminists and gay men in San Francisco', in *An Invitation to Geography*, Lanegran, D. and Palm, R. (eds), 2nd edn, New York: McGraw Hill, 78–88.

Lyotard, J. (1984) *The Postmodern Condition: A Report on Knowledge*, Minneapolis: University of Minnesota Press.

MacDonald, G.M. (1995) 'Indonesia's *Medan Merdeka*: national identity and the built environment', *Antipode*, 27(3), 270–93.

Mackenzie, S. (1989) 'Women in the city', in *New Models in Geography: The Political-Economy Perspective*, Peet, R. and Thrift, N. (eds), London: Unwin Hyman, 109–26.

Maddox, G. (1985) 'Hogan's feature film debut as Dundee', *Encore*, 20 June, 7.

Maddox, G. (1988) 'Having an ace up your sleeve', *Encore*, 29 August, 21.

Maitland, B. (1990) *The New Architecture of the Retail Mall*, New York: Van Nostrand Reinhold.

Malcolm, X. (1966) *Autobiography of Malcolm X*, New York: Grove.

Malm, K. and Wallis, R. (1993) *Media Policy and Music Activity*, London: Routledge.

Mann, S. (2001) 'Children walk blood-soaked path of history', *The Sydney Morning Herald*, 8–9 September, 21.

Marcus, G. and Fisher, M. (1986) *Anthropology as Cultural Critique*, Chicago: University of Chicago Press.

Marx, K. (1957) *Das Kapital* (*Capital*), translated from the 4th German edn by E. and C. Paul, London: J.M. Dent.

Massey, D. (1992) 'A place called home?', *New Formations*, 17 (Summer), 3–15.

Master Plan Report of Survey (1955), Colony of Singapore: Government Printers.

McDonald's Corporation (2001a) *McDonald's History . . . Yesterday and Today*, at: http://www.mcdonalds.com/countries/usa/corporate/info/studentkit/index.html (accessed 30 August 2001).

McDonald's Corporation (2001b) *Financial Report, 2000–2001*, Oak Brook: McDonald's Corporation.

McDonald-Walker, S. (1998) 'Fighting the legacy: British bikers in the 1990s', *Sociology*, 32(2), 379–96.

McDowell, L. (1994) 'The transformation of cultural geography', in *Human Geography: Society, Space and Social Science*, Gregory, D., Martin, R. and Smith, G. (eds), Basingstoke: Macmillan Press, 146–73.

McDowell, L.M. and Court, G. (1994) 'Missing subjects: gender, power, and sexuality in merchant banking', *Economic Geography*, 70(3), 229–52.

McGregor, F. (1996) 'I am not a lesbian', *Meanjin*, 55(1), 31–40.

McLeay, C.R. (1998) *The Circuit of Popular Music: Production, Consumption, Globalisation*, PhD Thesis, Sydney: Macquarie University.

McManus, P. and Pritchard, B. (2000) 'Geography and the emergence of rural and regional Australia', *Australian Geographer*, 31(3), 383–91.

Meadows, M. and Oldham, C. (1991) 'Racism and the dominant ideology: Aborigines, television news and the Bicentenary', *Media Information Australia*, 60, 30–40.

Mee, K. and Dowling, R. (2000) 'Working class masculinity, suburbia and resistance: David Caesar's Idiot Box', in 'Proceedings of the Habitus Conference', Perth: Curtin University of Techology.

Mee, K. and Dowling R. (2001) 'Reflections on representation: critical reactions to Idiot Box', *Research Papers of the Centre for Urban and Regional Studies*, No. 04, Newcastle: University of Newcastle.

Meinig, D.W. (1979a) 'Introduction', in *The Interpretation of Ordinary Landscapes: Geographical Essays*, Meinig, D.W. (ed.), New York: Oxford University Press, 1–7.

Meinig, D.W. (1979b) 'The beholding eye: ten versions of the same scene', in *The Interpretation of Ordinary Landscapes: Geographical Essays*, Meinig, D.W. (ed.), New York: Oxford University Press, 33–48.

Melucci, A. (1989) *Nomads of the Present: Social Movements and Individual Needs in Contemporary Society*, London: Hutchinson Radius.

Meyrowitz, J. (1985) *No Sense of Place*, Oxford: Oxford University Press.

Mishkind, M., Rodin, J., Silberstein, L., and Striegel-Moore, R. (1987) 'The embodiment of masculinity', in *Changing Men: New Directions in Research on Men and Masculinity*, Kimmel, S. (ed.), Newbury Park: Sage, 37–52.

Mitchell, D. (1996) 'Introduction: public space and the city', *Urban Geography*, 17, 127–31.

Monk, J. (1992) 'Gender in the landscape: expressions of power and meaning', in *Inventing Places: Studies in Cultural Geography*, Anderson, K. and Gale, F. (eds), Melbourne: Longman Cheshire, 123–38.

Moriarty, D.J. (1999) 'When a meal at McDonald's isn't so happy', *National Catholic Reporter*, 35(11), 21.

Morris, M. (1982) 'Sydney Tower', *Island Magazine*, 9/10, 53–61.

Morris, M. (1988) 'Tooth and claw: tales of survival and *Crocodile Dundee*', in *Universal Abandon? The Politics of Postmodernism*, Ross, A. (ed.), Minneapolis: University of Minnesota Press, 105–27.

Mullins, P. (1993) 'Tourism urbanisation', *International Journal of Urban and Regional Research*, 15(3), 326–42.

Mumford, L. (1961) *The City in History: Its Origins, its Transformations, and its Prospects*, London: Penguin.

Murphree, D.W., Wright, S.A. and Ebaugh, H.R. (1996) 'Toxic waste siting and community resistance: how cooptation of local citizen opposition failed', *Sociological Perspectives*, 39(4), 447–63.

Murphy, P. and Watson, S. (1997) *Surface City*, Annandale: Pluto Press.

Namaste, K. (1996) 'Genderbashing: sexuality, gender, and the regulation of public space', *Environment and Planning D: Society and Space*, 14(2), 221–40.

Naylor, S.K. and Ryan, J.R. (1998) 'Ethnicity and cultural landscapes: mosques, gurdwaras and mandirs in England and Wales', paper presented at the 'Religion and Locality Conference', University of Leeds, 8–10 September.

Not the Singapore Song Book (1993) Singapore: Hotspot Books.

O'Regan, T. (1988) '"Fair dinkum fillums": the *Crocodile Dundee* phenomenon', in *The Imaginary Industry: Australian Film in the 1980s*, Dermody, S. and Jacka, E. (eds), Sydney: Australian Film, Television and Radio School, 155–75.

Oakley, A. (1986) *From Here to Maternity: Becoming a Mother*, 2nd edn, Harmondsworth: Penguin.

Office of Multicultural Affairs (1989) *National Agenda for a Multicultural Australia . . . Sharing Our Future*, Canberra: Australian Government Publishing Service.

Oliver, H. (1990) *The Politics of Disablement*, London: Macmillan.

Orbach, S. (1982) *Fat is a Feminist Issue*, New York: Berkeley.

Owen, K.A. (2002) 'The Sydney 2000 Olympics and urban entrepreneurialism: local variations in urban governance', *Australian Geographical Studies*, 40(3), 323–36.

Pain, R. (1991) 'Space, sexual violence and social control: integrating geographical and feminist analyses of women's fear of crime', *Progress in Human Geography*, 15, 415–31.

Park, C.C. (1994) *Sacred Worlds: An Introduction to Geography and Religion*, London: Routledge.

Parker, S., Nichter, M., Vuckovic, N., Sims, C. and Ritenbaugh, C. (1995) 'Body image and weight concerns among African American and white adolescent females: differences that make a difference', *Human Organization*, 54(2), 103–14.

Peake, L. (1993) '"Race" and sexuality: challenging the patriarchal structuring of urban social space', *Environment and Planning D: Society and Space*, 11(4), 415–32.

Pearson, J. (1982) 'Conflict, stereotypes and masculinity in Australian and New Zealand surfing', *Australian and New Zealand Journal of Sociology*, 18(2), 117–35.

Perry, M., Kong, L. and Yeoh, B. (1997) *Singapore: A Developmental City-State*, Chichester: John Wiley.

Phua, L. (1998) *Negotiating Power Relations in Everyday Spaces: The Politics of the Pregnant Body in Singapore*, MA Thesis, Department of Geography, National University of Singapore.

Pickett, S. (2001) 'Heaven on a plate', *Vive*, 20 (Spring), 106–9.

Porteous, J.D. (1988) 'Topocide: the annihilation of place', in *Qualitative Methods in Human Geography*, Eyles, J. and Smith, D.M. (eds), Cambridge: Polity Press, 75–93.

Posel, D. (1988) 'Providing for the legitimate labour requirements of employers: secondary industry, commerce and the state in South Africa in the 1950s and early 1960s', in *Organisation and Economic Change: Southern African Studies*, Mabin, A. (ed.), Johannesburg: Ravan Press, 5, 199–220.

Potteiger, M. and Purinton, J. (1998) *Landscape Narratives: Design Practices for Telling Stories*, New York: John Wiley.

Powell, D. (1993) *Out West: Perceptions of Sydney's Western Suburbs*, Sydney: Allen & Unwin.

Powell, J.M. (1988) 'Geographical education and its Australian heritage', *Australian Geographer*, 26(1), 214–30.

Power, A. (1996) 'Area-based poverty and resident empowerment', *Urban Studies*, 33(9), 1535–64.

Powers, L.A. (1996) 'What happened to the graffiti art movement?', *Journal of Popular Culture*, 29 (Spring), 137–42.

Pred, A. (1992) 'Capitalism, crises, and cultures II: notes on local transformation and everyday cultural struggles', in *Reworking Modernity: Capitalisms and Symbolic Discontent*, Pred, A. and Watts, M. (eds), New Brunswick: Rutgers University Press, 106–17.

Price, M. and Lewis, M. (1993) 'The reinvention of cultural geography', *Annals of the Association of American Geographers*, 83(1), 1–17.

Pulvirenti, M. (1997) 'Unwrapping the parcel: an examination of culture through Italian Australian home ownership', *Australian Geographical Studies*, 35(1), 32–9.

Rajkowski, P. (1987) *In the Tracks of the Camelmen*, North Ryde: Angus & Robertson.

Rapoport, A. (1984) 'Culture and the urban order' in *The City in Cultural Context*, Agnew, J.A., Mercer, J. and Sopher, D.E. (eds), Boston: Allen & Unwin, 50–75.

Relph, E. (1991) 'Post-modern geographie', *Canadian Geographer*, 35, 98–105.

Ripe, C. (1994) 'Advance Australian fare, folks', *The Weekend Review*, August, 27–28.

Ripe, C. (1996) 'Bush-food providers cultivate the big time', *The Weekend Australian*, March, 16–17.

Ritzer, G. (1992) *McDonaldization of Society*, Thousand Oaks: Pine Forge.

Robinson, J. (1997) 'The geopolitics of South African cities: states, citizens, territory', *Political Geography*, 16(5), 365–86.

Robinson, T. (1996) 'Inner-city innovator: the non-profit community development corporation', *Urban Studies*, 33(9), 1647–70.

Roddick, N. (1986) 'Crocs away', *Cinema Papers*, 58 (July), 40.

Rosenthal, F. (1975) *The Classical Heritage in Islam*, translated by E. and J. Marmorstein, Berkeley: University of California Press.

Ross, C.E. and Jang, S.J. (2000) 'Neighbourhood disorder, fear, and mistrust: the buffering role of social ties with neighbours', *American Journal of Community Psychology*, 28(4), 401–20.

Ross, C.E., Pribesh, S. and Mirowsky, J. (2001) 'Powerlessness and the amplification of threat: neighbourhood disadvantage, disorder and mistrust', *American Sociological Review*, 66, 568–91.

Rostow, W. (1960) *The Stages of Economic Growth: A Non-Communist Manifesto*, London: Cambridge University Press.

Rothenberg, T. (1995) '"And she told two friends": lesbians creating urban social space', in *Mapping Desire: Geographies of Sexualities*, Bell, D. and Valentine, G. (eds), London: Routledge, 165–81.

Rowe, D. (1994) 'Accommodating bodies: celebrity, sexuality and "Tragic Magic"', *Journal of Sport and Social Issues*, 18(1), 7–26.

Rowntree, L.B. (1988) 'Orthodoxy and new directions: cultural/humanistic geography', *Progress in Human Geography*, 12(4), 575–86.

Rowse, T. and Moran, A. (1984) '"Peculiarly Australian" – the political construction of cultural identity', in *Australian Society: Introductory Essays*, Encel, S., Berry, M., Bryson, L., de Lepervanche, M., Rowse, T. and Moran, A. (eds), 4th edn, Melbourne: Longman Cheshire, 229–77.

Royal Commission into Aboriginal Deaths in Custody (RCIADIC) (1991) *National Report Volume 2: The Underlying Issues which Explain the Disproportionate Number of Aboriginal People in Custody*, Canberra: Australian Government Publishing Service.

Rubenstein, J.M. and Bacon, R.S. (1983) *The Cultural Landscape: An Introduction to Human Geography*, Saint Paul: West Publishing.

Ruddick, S. (1996) *Young and Homeless in Hollywood: Mapping Social Identities*, London: Routledge.

Russell, D. (1991) 'What's wrong with brass bands? Cultural change and the band movement, 1918–*c*1964', in *Bands: The Brass Band Movement in the 19th and 20th centuries*, Herbert, T. (ed.), Milton Keynes: Open University Press, 57–101.

Ryan, J. and Thomas, F. (1978) *The Politics of Mental Handicap*, London: Free Association.

Rydell, R. (1984) *All the World's a Fair*, Chicago: University of Chicago Press.

Said, E.W. (1978) *Orientalism*, New York: Vintage Books.

Sanders, W. (1993) 'Aboriginal housing', in *Housing Australia*, Paris, C. (ed.), Melbourne: Macmillan, 212–27.

Sauer, C.O. (1969) 'The morphology of landscape', in *Land and Life: A Selection from Writings of Carl Ortwin Sauer*, Leighley, J. (ed.), Berkeley: University of California Press, 315–50.

Saunders, P. (1984) 'The Canberra tea party: bureaucracy, pluralism and corporatism in the administration of the Australian Capital Territory', in *Conflict and Development*, Williams, P. (ed.), Sydney: Allen & Unwin, 51–75.

Savage, V.R. and Kong, L. (1993) 'Urban constraints, political imperatives: environmental "design" in Singapore', *Landscape and Urban Planning*, 25(1–2), 37–52.

Schiller, N.G. (1994) 'Introducing identities: global studies in culture and power', *Identities*, 1, 1–6.

Scott, J.C. (1985) *Weapons of the Weak: Everyday Forms of Peasant Resistance*, New Haven: Yale University Press.

Seah, C.M. (1989) 'National security', in *Management of Success: The Moulding of Modern Singapore*, Sandhu, K.S. and Wheatley, P. (eds), Singapore: Institute of Southeast Asian Studies, 949–62.

Sedgwick, E.K. (1990) *Epistemology of the Closet*, Berkeley: University of California Press.

Sedgwick, E.K. (1993) *Tendencies*, Durham: Duke University Press.

Seebohm, K. (1994) 'The nature and meaning of the Sydney Mardi Gras in a landscape of inscribed social relations' in Aldrich, R. and Wotherspoon, G. (eds) *Gay Perspectives II: More Essays in Australian Gay Culture*, Sydney: University of Sydney Press, 193–222.

Semple, E. (1911) *Influences of Geographic Environment: On the Basis of Ratzel's System of Anthropo-geography*, New York: Henry Holt.

Sharaf, A.T. (1963) *A Short History of Geographical Discovery*, Alexandria: M. Zaki El Mahdy.

Shields, R. (1991) *Places on the Margin: Alternative Geographies of Modernity*, London: Routledge.

Shirlow, P. (1999) *Fear, Mobility and Living in the Ardoyne and Upper Ardoyne Communities*, Belfast: North Belfast Partnership Board.

Short, J.R., Benton, L.M., Luce, W.B. and Walton, J. (1993) 'Reconstructing the image of an industrial city', *Annals of the Association of American Geographers*, 83(2), 207–24.

Sibley, D. (1988) 'Survey 13: purification of space', *Environment and Planning D: Society and Space*, 6, 409–21.

Sibley, D. (1995) *Geographies of Exclusion: Society and Difference in the West*, London: Routledge.

Sibley, D. (1999) 'Outsiders in society and space', in *Cultural Geographies*, Anderson, K. and Gale, F. (eds), 2nd edn, London: Addison Wesley Longman, 135–51.

Simkins, C. (1981) 'Agricultural production in the African reserves of South Africa, 1918–1969', *Journal of South African Studies*, 7, 176–95.

Sing Singapore: A Celebration in Song, National Day 1988 (1988), Singapore: F & N Sarsi.

Singapore: Official Guide (1991) Singapore: Singapore Tourist Promotion Board.

Smith, A.S. (1990) 'Towards a global culture?', in *Global Culture: Nationalism, Globalization and Modernity*, M. Featherstone (ed.), London: Sage, 171–92.

Smith, J.R. (1913) *Industrial and Commercial Geography*, New York: Henry Holt.

Smith, M. (1996) 'The empire filters back: consumption, production, and the politics of Starbucks Coffee', *Urban Geography*, 17(6), 502–24.

Smith, N. (1979) 'Gentrification and capital: practice and ideology in Society Hill', *Antipode*, 11(3), 24–35.

Smith, S.J. (1994) 'Soundscape', *Area*, 26, 232–40.

Soja, E.W. (1993) 'Postmodern cities', transcript of conference address at the 'Postmodern Cities Conference', Department of Urban and Regional Planning, University of Sydney.

Soja, E.W. (1995) 'Postmodern urbanization', in *Postmodern Cities and Spaces*, Watson, S. and Gibson, K. (eds), Cambridge: Blackwell.

Sopher, D.E. (1967) *Geography of Religions*, New Jersey: Prentice-Hall.

Sorkin, M. (ed.) (1992) *Variations on a Theme Park: The New American City and the End of Public Space*, New York: Hill & Wang.

Spirn, A.W. (1998) *The Language of Landscape*, New Haven: Yale University Press.

Spooner, R. (1996) 'Contested representations: black women and the St Paul's Carnival', *Gender, Place and Culture*, 3(2), 187–203.

Stam, R. (1988) 'Mikhail Bakhtin and the Left cultural critique', in *Postmodernism and its Discontents*, Kaplan, E.A. (ed.), London: Verso, 116–45.

Stock, B. (1986) 'Texts, readers, and enacted narratives', *Visible Language*, 20(3), 294–301.

Stratford, E. (1999) 'Australian cultural geographies', in *Australian Cultural Geographies*, Stratford, E. (ed.), South Melbourne: Oxford University Press, 1–10.

Street, J. (1993) 'Local differences? Popular music and the local state', *Popular Music*, 12, 43–55.

Taylor, G. (1940) *Australia: A Study of Warm Environments and their Effect on British Settlement*, London: Methuen.

Taylor, G. (1949) *Environment, Race and Migration*, 3rd edn, Toronto: University of Toronto Press.

Teather, E.K. (1998) 'Themes from complex landscapes: Chinese cemeteries and columbaria in urban Hong Kong', *Australian Geographical Studies*, 36(1), 21–36.

Thompson, E.P. (1968) *The Making of the English Working Class*, Harmondsworth: Penguin.

Thompson, J.B. (1981) *Critical Hermeneutics*, Cambridge: Cambridge University Press.

Thrift, N. (1994) 'Review of *Writing Worlds*', *Antipode*, 26(1), 109–11.

Turner, V. (1974a) *The Ritual Process*, Harmondsworth: Penguin.

Turner, V. (1974b) *Dramas, Fields and Metaphors*, Ithaca: Cornell University Press.

UNESCO (1983) 'Racism, science and pseudo-science', proceedings of the 'Symposium to Examine Pseudo-Scientific Theories Invoked to Justify Racism and Racial Discrimination', Athens, 30 March–3 April 1981, Paris: UNESCO.

URA (1983) *A Pictorial Chronology of the Sale of Sites Programme for Private Development*, Singapore: Urban Redevelopment Authority.

Urban Redevelopment Authority Annual Report (1986/87).

Valentine, G. (1989) 'The geography of women's fear', *Area*, 21, 385–90.

Valentine, G. (1990) 'Women's fear and the design of public spaces', *Built Environment*, 16, 288–303.

Valentine, G. (1992) 'Images of danger: women's sources of information about the spatial distribution of male violence', *Area*, 24, 22–9.

Valentine, G. (1993a) '(Hetero)sexing space: lesbian perceptions and experiences of everyday spaces', *Environment and Planning D: Society and Space*, 11(4), 395–413.

Valentine, G. (1993b) 'Negotiating and managing multiple sexual identities: lesbian time-space strategies', *Transactions of the Institute of British Geographers*, 18(2), 237–48.

Valentine, G. (1993c) 'Desperately seeking Susan: a geography of lesbian friendships', *Area*, 25(2), 109–16.

Valentine, G. (1995) 'Out and about: geographies of lesbian landscapes', *International Journal of Urban and Regional Research*, 19(1), 96–111.

van Gennep, A. (1960) *The Rites of Passage*, translated by M.B. Vizedom and G.L. Caffee, London: Routledge & Kegan Paul (first published in 1909, *Les Rites de Passage*, Noury.)

Vidal de la Blache, P. (1917) *France de l'Est*, Paris.

Waitt, G., McGuirk, P., Dunn, K.M., Hartig, K.V. and Burnley, I.H. (2000) *Introducing Human Geography: Globalisation, Difference and Inequality*, Sydney: Pearson Education.

Walker, L. (1995) 'More than just skin deep: fem(me)ininity and the subversion of identity', *Gender, Place and Culture*, 2(1), 71–6.

Wallis, R. and Malm, K. (1984) *Big Sounds from Small Peoples*, London: Constable.

Wallis, R. and Malm, K. (1987) 'The international music industry and transcultural communication', in *Popular Music and Communication*, Lull, J. (ed.), Newbury Park: Sage, 112–37.

Walton, J.R. (1995) 'How real(ist) can you get?', *The Professional Geographer*, 47(1), 61–5.

Walton, J.R. (1996) 'Bridging the divide – a reply to Mitchell and Peet', *The Professional Geographer*, 48(1), 98–100.

Ward, J. (2001) 'The writing's on the wall, and we want it off', *American City and Country*, June, 4.

Warner, M. (1985) *Monuments and Maidens: The Allegory of the Female Form*, London: Weidenfeld & Nicolson.

Warren, S. (1994) 'Disneyfication of the metropolis: popular resistance in Seattle', *Journal of Urban Affairs*, 16(2), 89–107.

Warren, S. (1996) 'Popular cultural practices in the "postmodern city"', *Urban Geography*, 17(6), 545–67.

Watts, M.J. (1991) 'Mapping meaning, denoting difference, imagining identity: dialectical images and postmodern geographies', *Geografiska Annaler*, 73B(1), 7–15.

Weightman, B.A. (1980) 'Gay bars as private places', *Landscape*, 24(1), 9–16.

Weightman, B.A. (1981) 'Commentary: towards a geography of the gay community', *Journal of Cultural Geography*, 1, 106–12.

Weightman, B.A. (1988) 'Sign geography', *Journal of Cultural Geography*, 9(1), 53–70.

Wekerle, G., Peterson, R. and Morley, D. (1980) 'Introduction', in *New Space for Women*, Wekerle, G., Peterson, R. and Morley, D. (eds), Boulder: Westview Press, 1–34.

Whatmore, S. and Boucher, S. (1993) 'Bargaining with nature: the discourse and practice of "environmental planning gain"', *Transactions of the Institute of British Geographers*, 18(2), 166–78.

White, P.E. (1984) *The Western European City*, Harlow: Longman.

White, P.E. and Winchester, H.P.M. (1991) 'The poor in the inner city: stability and change in two Parisian neighbourhoods', *Urban Geography*, 12(1), 35–54.

Williams, J.J. (2000) 'South Africa: urban transformation', *Cities*, 17(3), 167–83.

Williams, R. (1958) *Culture and Society, 1780–1950*, London: Chatto & Windus.

Willis, P.E. (1978) *Profane Culture*, London: Routledge & Kegan Paul.

Winchester, H.P.M. (1990) 'Women and children last: the poverty and marginalization of one-parent families', *Transactions of the Institute of British Geographers*, 15(1), 70–86.

Winchester, H.P.M. (1992) 'The construction and deconstruction of women's roles in the urban landscape', in *Inventing Places: Studies in Cultural Geography*, Anderson, K. and Gale, F. (eds), Melbourne: Longman Cheshire, 139–56.

Winchester, H.P.M. (1996) 'Ethical issues in interviewing as a research method in human geography', *Australian Geographer*, 2(1), 117–31.

Winchester, H.P.M. and Costello, L.N. (1995) 'Living on the street: social organization and gender relations of Australian street kids', *Environment and Planning D: Society and Space*, 13(3), 329–48.

Winchester, H.P.M. and White, P.E. (1988) 'The location of marginalised groups in the inner city', *Environment and Planning D: Society and Space*, 6(1), 37–54.

Winchester, H.P.M., McGuirk, P.M. and Everett, K. (1999) 'Schoolies Week as a rite of passage', in *Embodied Geographies: Spaces, Bodies and Rites of Passage*, Teather, E.K. (ed.), London: Routledge, 59–77.

Wolch, J.R. (1995) 'Inside/outside: the dialectics of homelessness', in *Populations at Risk in America: Vulnerable Groups at the End of the Twentieth Century*, Demko, G.J. and Jackson, M.C. (eds), Boulder: Westview Press, 77–90.

Wolf, N. (1990) *The Beauty Myth*, London: Chatto and Windus.

Wolpse, H. (1974) 'Capitalism and cheap labour power in South Africa: from segregation to apartheid', *Economy and Society*, 1, 425–56.

Wood, D. (1995) 'Conserved to death: are tropical forests being overprotected from people?', *Land Use Policy*, 12(2), 115–35.

Yang, C.K. (1961) *Religion in Chinese Society*, Berkeley: University of California Press.

Yeoh, B.S.A. (1999) 'The body after death: place, tradition and the nation-state in Singapore', in *Embodied Geographies: Spaces, Bodies and Rites of Passage*, Teather, E.K. (ed.), London: Routledge, 240–55.

Yeoh, B.S.A. and Kong, L. (1994) 'Reading landscape meanings: state constructions and lived experiences in Singapore's Chinatown', *Habitat International*, 18(4), 17–35.

Young, E. (1993) 'Hunter-gatherer concepts of land and its ownership in remote Australia and North America', in *Inventing Places: Studies in Cultural Geography*, Anderson, K. and Gale, F. (eds), Melbourne: Longman Cheshire, 255–72.

Young, I.M. (1990) *Justice and the Politics of Difference*, Princeton: Princeton University Press.

Zelinsky, W. (1973) *The Cultural Geography of the United States*, Englewood Cliffs: Prentice-Hall.

Zonn, L.E. (1984) 'Images of place: a geography of the media', Proceedings of the Royal Geographical Society of Australasia, 84, 35–45.

Zukin, S. (1991) *Landscapes of Power: From Detroit to Disney World*, Berkeley: University of California Press.

Index

Rip Curl, 165
rites of passage, 149–50, 152, 153, 155
Rodeo Drive (USA), 78
Romper Stomper (film), 44, 47
Ruhr (Germany), 133

Sahara (Africa), 36
Sakran discourse, 97
San Antonio (USA), 91
San Francisco (USA), 105, 139, 154
San Marco, Venice, 72
Sauer, Carl, 15, 16, 17, 18, 19, 22, 23, 174
Schoolies Week (Australia), 152–3
Scotland, 55, 133
Seattle (USA), 91, 93, 94
Seattle Center (USA), 93, 94
self, the, 8, 31, 47, 111, 176, 177
Second Empire (Napoleon), 94, 98
Semple, Ellen, 11, 12, 174
Senator Hotel (USA), 139
Seventh-day Adventists, 170
sexism, 5, 178
sexuality, 5
    bisexuals, 88
    eroticism, 160
    gays, 4, 21, 88, 105, 118, 120, 130, 154,
        155, 166, 171, 177
    geographical analyses of, 20–1
    hetero-centric dialogue, 87
    heteronormativity, 87, 154
    heterosexuality, 9, 67, 87, 89, 107, 155,
        158, 171, 172, 177
        hegemonic, 120
        naturalness of, 88
        norm of, 132
        signifiers of, 88
    homoeroticism, 160
    homophobia, 5, 21, 120, 154, 171, 177,
        178
    homosexuality, 21, 89, 105, 158, 159, 177
    lesbians/lesbianism, 4, 21, 88, 105, 118,
        120, 130, 154, 155, 166, 171, 177
    sex industry (*see* marginal, prostitutes/
        prostitution)
    transexuals, 103
    transgender, 88
    transvestites, 103
Shine (film), 44, 47
shopping malls, 64, 77–81, 93, 98, 161,
    163, 178
Siberia, 2
Sikh community, 113, 175
Singapore, 8, 13, 22, 30, 40, 42, 51, 52, 53,
    54, 55, 57, 62, 65, 68, 80, 98, 113,
    120, 143, 151, 161, 162, 175

Singapore River, 68, 69
Singlish (Singapore English), 39–40
Sinkiang (China), 2
Skid Row, Los Angeles (USA), 108
skyscrapers, 75, 76, 98, 128
Smoky Mountain, Manila (Philippines),
    128, 130, 131
social construction theory/approach, 7,
    30–2, 149, 156, 175
social Darwinism, 92
South Africa, 84, 85, 89, 91, 98, 176
Sovereign Islands, Queensland (Australia),
    82–3
Sri Lanka (ancient Kandy), 8, 97, 101, 122
Starbucks Coffee, 37, 38
state
    and capital, 91–4
    and religion, 94, 97
Statue of Liberty (USA), 89
STB (Singapore Tourism Board), 138
streetscapes, 36
Stockholm (Sweden), 106
Stockholm syndrome, 114
Strictly Ballroom (film), 44, 47
street
    life, 101
    names, 70, 74, 75, 101, 106
    queered, 178
    signs, 8, 175
    streetscapes, 110, 142
Sum of Us, The (film), 44, 47
superorganicism, 19, 24
superorganism, 22
surfers, 163–5, 175
Surfers' Paradise, Queensland (Australia), 82,
    175
surfies, 159, 163–5, 175
Swan Brewery, Perth (Australia), 11
Sydney (Australia), 8, 25, 27, 34, 45, 95,
    105, 110, 120, 128, 129, 133, 142,
    143, 144, 154, 176, 177
Sydney City Council, 95
Sydney Harbour, 134
Sydney Strategic Plan (1971), 95
Sydney 2000 Olympic Games, 135
Sydney Tower, 95, 96
Syracuse (USA), 133

Taliban government of Afghanistan, 30,
    156, 178
Tamils, 122
Tamil Tigers (Liberation Tigers of Tamil
    Eelam [*aka* LTTE]), 122
Tanzania, 22
Target, 159